Microelectronics and Optoelectronics Technology

MICROELECTRONICS AND OPTOELECTRONICS TECHNOLOGY

By

Saradindu Panda
M.Tech., Ph.D (Persuing)
Senior Lecturer, ECE Department
Narula Institute of Technology
Agarpara, Kolkata (West Bengal)

UNIVERSITY SCIENCE PRESS
(An Imprint of Laxmi Publications Pvt. Ltd.)
An ISO 9001:2015 Company
BENGALURU • CHENNAI • GUWAHATI • HYDERABAD
KOCHI • KOLKATA • LUCKNOW • RANCHI
NEW DELHI

Printed and bound in India

First Edition : 2009; *Edition* : 2025

ISBN : 978-81-31807-38-5

Branches

Branch	Telephone
Bengaluru	080-26 75 69 30
Chennai	044-24 34 47 26
Guwahati	0361-254 36 69
Hyderabad	040-27 55 53 83
Kochi	0484-405 13 03
Kolkata	033-40 04 77 79
Lucknow	0522-430 36 13
Ranchi	0651-224 24 64

Published in India by

UNIVERSITY SCIENCE PRESS

(An Imprint of Laxmi Publications Pvt. Ltd.)

An ISO 9001:2015 Company

113, GOLDEN HOUSE, GURUDWARA ROAD, DARYAGANJ,
NEW DELHI - 110002, INDIA

Corporate Office : B-33, Sector-5, Noida,
Gautambuddha Nagar (U.P.) - 201 301, INDIA

Telephone : +91-120-44 46 900

www.laxmipublications.com

info@laxmipublications.com

CONTENTS

PREFACE

The discipline of Microelectronics has played a fundamental role in shaping the electronics industry, as well as related industries that rely on electronic components and sub-systems. In a realm where changes happen frequently and dramatically, the constant themes that have persisted are miniaturization, increased speed, reduced power consumption, and reduced cost. These effects have resulted in an increased demand for microelectronics in all the sectors of consumer, industrial, and military products. Advancements in manufacturing have enabled these devices to be produced in very high volumes, thereby reducing the cost per device. In turn, the lower cost fuels future demand which pushes the industry for further miniaturization and higher volume manufacturing.

The market demand for microelectronics has evolved from largely a military-driven demand to one that is now largely consumer-driven. Consequently, the device features have also been targeted at consumer needs, such as low power, low cost, and mass market applications, rather than military needs, such as meeting military specifications for reliability and packaging, specialized applications, and the resulting high cost of such devices. The goal is to enable the end user of the devices to perform complex tasks in a more efficient manner than what was previously possible.

This book is mainly addressed to the final year undergraduate and beginning graduate students in physics, materials science, and several kinds of engineerings (electrical, materials, etc.) with the objective in mind that it could be used in one semester. Alternatively, the book can be of interest to scientists and practicing engineers who want to know about the fundamental aspects Micro and Optoelectronics devices. My intention has been to write an introductory book on Micro and Optoelectronics devices that starting with the physics of low-dimensional semiconductors and quantum heterojunctions would build up to the treatment of those new electronic, transport, and optical properties, which arise as a consequence of both energy quantization of electrons in potential wells.

The main objectives are:

(*i*) The extension of the book has been limited so that it could be taught in one semester.

(*ii*) I have tried to make as clear as I could the new physical concepts and properties presented in the book.

(*iii*) It includes the fundamental concept of different types contacts (Ohmic and Non-ohmic), MOS Capacitors and MOSFET, the Concept of band bending, Concept of Heterojunction and their band structures, Charge storage devices, Power MOSFETs and different types of Optoelectronics Devices.

(*iv*) Most notably an important chapter has been written to reflect new applications for microfabrication processes, called Micro-Electro-Mechanical-Systems (MEMS). This existing area promises to open up many new areas for microfabrication. I welcome any suggestions and corrections for the improvement of the book.

— Author

ACKNOWLEDGEMENT

I am greatly thankful to my Professors, Prof. C.K. Sarkar and Prof. 'Chayanika Bose, Jadavpur University for their providing necessary class notes to complete the manuscript for the book. I am also thankful to all the staff and faculty and management of Narula Institute of Technology, Agarpara, West Bengal for their encouragement and providing necessary facilities to complete the manuscript for the book.

Finally, I express my thanks to my parents for their continuous support.

—Author

1

Semiconductor Fundamentals

1.1 INTRODUCTION

To understand the fundamental concepts of semiconductors, one must apply modern physics to solid materials. More specifically, we are interested in semiconductor crystals. Crystals are solid materials consisting of atoms, which are placed in a highly ordered structure called a lattice. Such a structure yields a periodic potential throughout the material, which results in some remarkable properties.

Two properties of crystals are of particular interest, since they are needed to calculate the current in a semiconductor. First, we need to know how many fixed and mobile charges are present in the material. Second, we need to understand the transport of the mobile carriers through the semiconductor.

In this chapter, we explain the concepts of energy bands, energy band gaps and the density of states in an energy band. We also show how the current in an almost filled band can more easily be analyzed using the concept of holes.

Two carrier transport mechanisms will be considered. The drift of carriers in an electric field and the diffusion of carriers due to a carrier density gradient will be discussed. Recombination mechanisms and the continuity equations are then combined into the diffusion equation. Finally, we present the drift-diffusion model, which combines all the essential elements discussed in this chapter.

1.2 ENERGY BANDS

Energy bands consisting of a large number of closely spaced energy levels exist in crystalline materials. The bands can be thought of as the collection of the individual energy levels of electrons surrounding each atom. The wave functions of the individual electrons, however, overlap with those of electrons confined to neighboring atoms. The

Pauli exclusion principle does not allow the electron energy levels to be the same so that one obtains a set of closely spaced energy levels, forming an energy band. The energy band model is crucial to any detailed treatment of semiconductor devices. It provides the framework needed to understand the concept of an energy bandgap and that of conduction in an almost filled band as described by the empty states.

In this section, we present the free electron model and the Kronig-Penney model. Then we discuss the energy bands of semiconductors and present a simplified band diagram. We also introduce the concept of holes and the effective mass.

1.2.1 Free Electron Model

The free electron model of metals has been used to explain the photo-electric effect. This model assumes that electrons are free to move within the metal but are confined to the metal by potential barriers as illustrated by Fig. 1.1. The minimum energy needed to extract an electron from the metal equals $q\Phi_m$, where Φ_M is the work-function. This model is frequently used when analyzing metals. However, this model does not work well for semiconductors since the effect of the periodic potential due to the atoms in the crystal has been ignored.

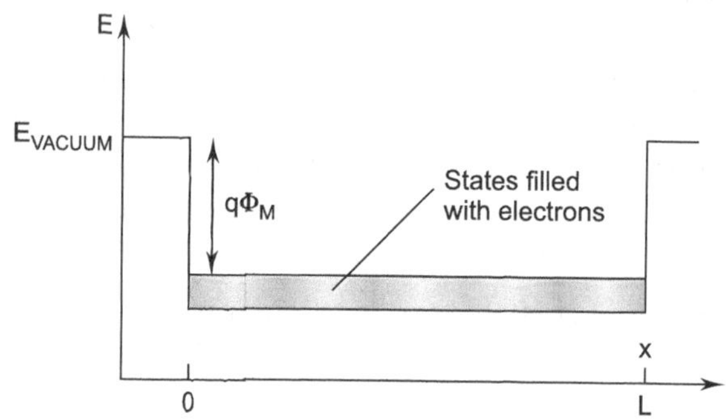

Fig. 1.1 *The free electron model of a metal*

1.2.2 Periodic Potentials

The anlaysis of periodic potentials is required to find the energy levels in a semiconductor. This requires the use of periodic wave functions, called Bloch functions which are beyond the scope of this text. The result of this analysis is that the energy levels are grouped in bands, separated by energy band gaps. The behavior of electrons at the bottom of such a band is similar to that of a free electron. However, the electrons are affected by the presence of the periodic potential. The combined effect of the periodic potential is included by adjusting the value of the electron mass. This mass will be referred to as the effective mass.

The effect of a periodic arrangement on the electron energy levels is illustrated by Fig. 1.2. Shown are the energy levels of electrons in a carbon crystal with the atoms arranged in a diamond lattice. These energy levels are plotted as a function of the lattice constant, a.

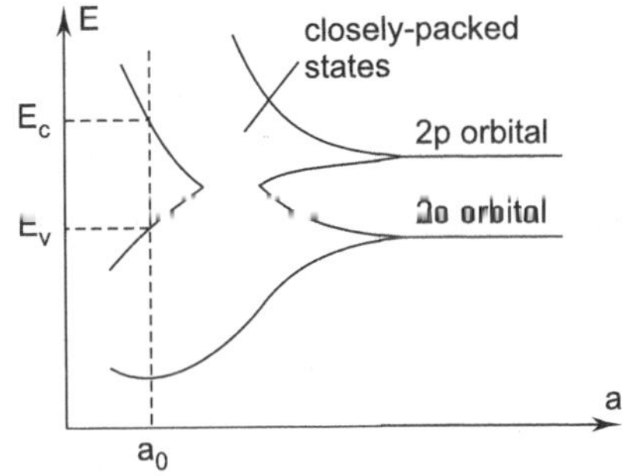

Fig. 1.2 *Energy bands for diamond versus lattice constant*

Isolated carbon atoms contain six electrons, which occupy the 1*s*, 2*s* and 2*p* orbital in pairs. The energy of an electron occupying the 2*s* and 2*p* orbital is indicated on the figure. The energy of the 1s orbital is not shown. As the lattice constant is reduced, there is an overlap of the electron wave functions occupying adjacent atoms. This leads to a splitting of the energy levels consistent with the Pauli exclusion principle. The splitting results in an energy band containing 2N states in the 2*s* band and 6N states in the 2*p* band, where N is the number of atoms in the crystal. A further reduction of the lattice constant causes the 2*s* and 2*p* energy bands to merge and split again into two bands containing 4N states each. At zero Kelvin, the lower band is completely filled with electrons and labeled as the valence band. The upper band is empty and labeled as the conduction band.

1.2.3 The Kronig-Penney Model

The Kronig-Penney model demonstrates that a simple one-dimensional periodic potential yields energy bands as well as energy band gaps. While it is an oversimplification of the three-dimensional potential and bandstructure in an actual semiconductor crystal, it is an instructive tool to demonstrate how the band structure can be calculated for a periodic potential, and how allowed and forbidden energies are obtained when solving the corresponding Schrodinger equation.

The potential assumed in the model is shown in the Fig. 1.3:

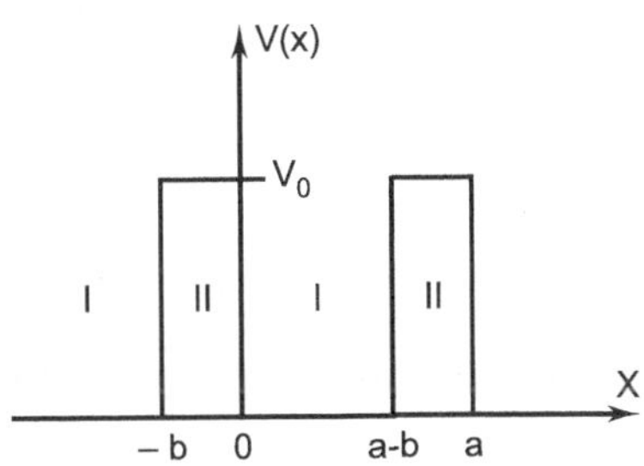

Fig. 1.3 *The periodic potential assumed in the Kronig-Penney model. The potential barriers (region I) with width, b, are spaced by a distance (region II), a-b, and repeated with a period, a*

The Kronig-Penney consists of an infinite series of rectangular barriers with potential height, V_0, and width, b, separated by a distance, a-b, resulting in a periodic potential with period, a. The analysis requires the use of Bloch functions, traveling wave solutions multiplied with a periodic function, which has the same periodicity as the potential.

Solutions for k and E are obtained when the following equation is satisfied:

$$\cos ka = F = \frac{\alpha^2 - \beta^2}{2\alpha\beta} \sinh \alpha b \sinh \beta (a-b) - \cosh \alpha b \cos \beta (a-b) \quad \text{... (1.1)}$$

where
$$\alpha = \frac{\sqrt{2m(V_0 - E)}}{\hbar} \text{ and } \beta = \sqrt{\frac{2mE}{\hbar}} \quad \text{... (1.2)}$$

This transcendental equation can be further simplified for the case where the barrier is a delta function with area, $V_0 b$, for which it becomes:

$$\cos ka = F = P \frac{\sin \beta a}{\beta a} + \cos \beta a \quad \text{... (1.3)}$$

with
$$\beta = \frac{\sqrt{2mE}}{\hbar} \text{ and } P = \frac{m V_0 ba}{\hbar^2} \quad \text{... (1.4)}$$

This equation can only be solved numerically. Solutions are only obtained if the function, F, is between – 1 and 1 since it has to equal cos (ka). The energy E, is plotted as function of $\frac{ka}{\pi}$ and the function F in Fig. 1.4.

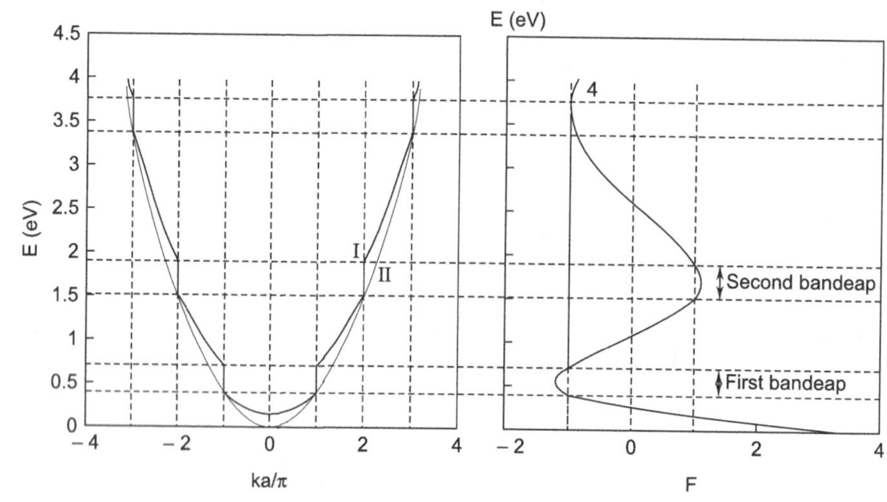

Fig. 1.4 *Graphical solution to the Kronig-Penney model for a = 1 nm and $V_0 b$ = 0.2 nm-eV. Shown the energy, E, versus $\frac{ka}{\pi}$ and energy E, versus F, which has to equal cos (ka), from which one can identify the allowed energies.*

The corresponding band structure is shown below (I curve) as well as the energy for a free electron (II curve). Three different forms are presented, namely the E (k) diagram, the E (k) diagram combined with the reduced-zone diagram as well as the reduced-zone diagram only.

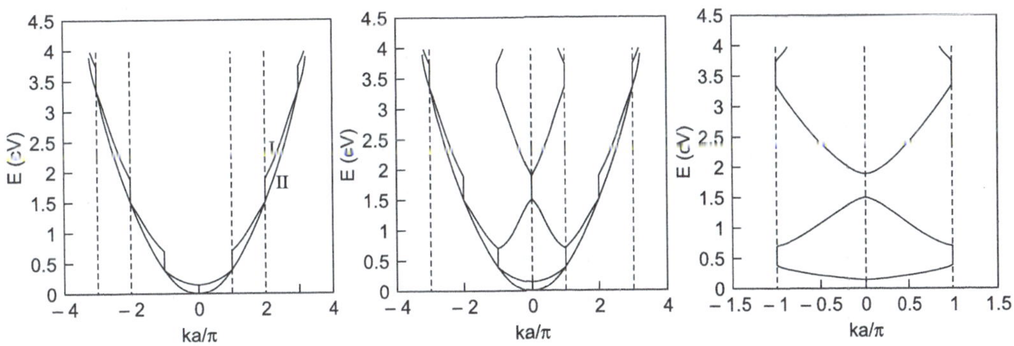

Fig. 1.5 *Energy versus $\frac{ka}{\pi}$ as presented in Fig. 1.4 (I curves) compared to that of a free electron (II curves). Shown are: (a) the E (k) diagram, (b) the E (k) diagram combined with the reduced-zone diagram and (c) the reduced-zone diagram only.*

From Fig. 1.5 (*a*) we observe the following: The E (*k*) relation resembles a parabola except that only specific ranges of energies are valid solutions to Schrodinger's equation and therefore are allowed, while others are not. The range of energies for which there is no solution is referred to as an energy band gap. The transitions between allowed and forbidden energies occur at non-zero integer multiples of $\frac{ka}{\pi}$. These correspond to local minima and maxima of the E (*k*) relation. The reduced-zone diagram shown in Fig. 1.5 (*c*) contains the first three bands and energy bandgaps. For instance the second energy bandgap occurs between 1.5 and 2 *e*V, between the band maximum of the second band and the band minimum of the third band.

1.3 ENERGY BANDS OF SEMICONDUCTORS

Complete energy band diagrams of semiconductors are very complex. However, most have features similar to that of the diamond crystal. In this section, we first take a closer look at the energy band diagrams of common semiconductors. We then present a simple diagram containing only the most important features and discuss the temperature and doping dependence of the energy bandgap.

1.3.1 Energy Band Diagrams of Common Semiconductors

The energy band diagrams of semiconductors are rather complex. The detailed energy band diagrams of germanium, silicon and gallium arsenide are shown in Fig. 1.6. The energy is plotted as a function of the wavenumber, *k*, along the main crystallographic directions in the crystal, since the band diagram depends on the direction in the crystal. The energy band diagrams contain multiple completely-filled and completely-empty bands. In addition, there are multiple partially-filled band.

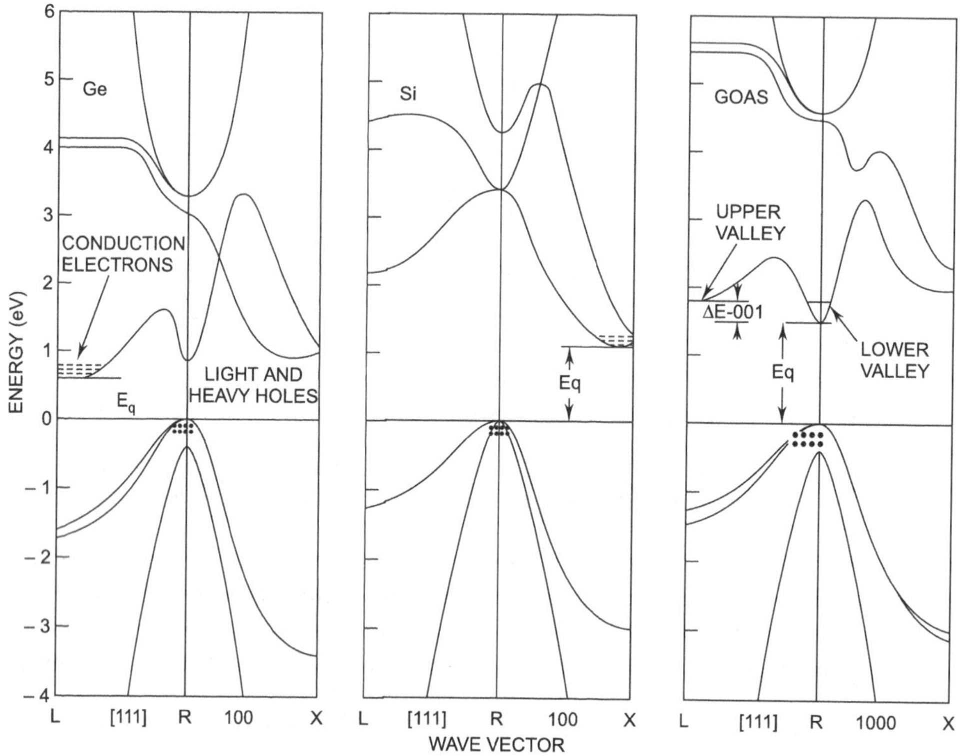

Fig. 1.6 *Energy band diagram of (a) germanium, (b) silicon and (c) gallium arsenide*

Fortunately, we can simplify the energy band diagram since only the electrons in the highest almost-filled band and the lowest almost-empty band dominate the behavior of the semiconductor. These bands are indicated on the Fig. by the + and – signs corresponding to the charge of the carriers in those bands.

1.3.2 Simple Energy Band Diagram of a Semiconductor

The energy band diagrams shown in the previous section are frequently simplified when analyzing semiconductor devices. Since the electronic properties of a semiconductor are dominated by the highest partially empty band and the lowest partially filled band, it is often sufficient to only consider those bands. This leads to a simplified energy band diagram for semiconductors as shown in Fig. 1.7:

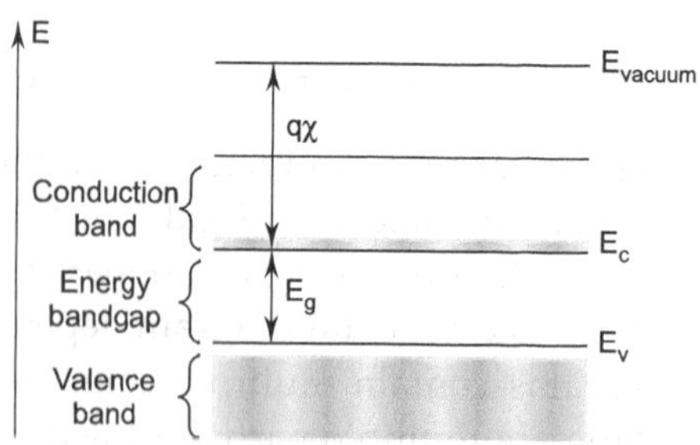

Fig. 1.7 *A simplified energy band diagram used to describe semiconductors. Shown are the valence and conduction band as indicated by the valence band edge, E_v, and the conduction band edge, E_c. The vacuum level, E_{vacuum}, and the electron affinity, χ, are also indicated on the figure*

The diagram identifies the almost-empty *conduction band* by a set of horizontal lines. The bottom line indicates the bottom edge of the conduction band and is labeled E_c. Similarly, the top of the valence band is indicated by a horizontal line labeled E_v. The energy bandgap, E_g, is located between the two bands. The distance between the conduction bandedge, E_c, and the energy of a free electron outside the crystal (called the vacuum level labeled E_{vacuum}) is quantified by the electron affinity, χ multiplied with the electronic charge q.

An important feature of an energy band diagram, which is not included on the simplified diagram, is whether the conduction band minimum and the valence band maximum occur at the same value for the wavenumber. If so, the energy bandgap is called direct. If not, the energy bandgap is called indirect. This distinction is of interest for optoelectronic devices since direct bandgap materials provide more efficient absorption and emission of light. For instance, the smallest bandgap of germanium and silicon is indirect, while gallium arsenide has a direct bandgap as can be seen on Fig. 1.7.

1.3.3 Temperature Dependence of the Energy Bandgap

The energy bandgap of semiconductors tends to decrease as the temperature is increased. This behavior can be understood if one considers that the interatomic spacing increases when the amplitude of the atomic vibrations increases due to the increased thermal energy. This effect is quantified by the linear expansion coefficient of a material. An increased interatomic spacing decreases the average potential seen by the electrons in the material, which in turn reduces the size of the energy bandgap. A direct modulation of the interatomic distance, such as by applying compressive (tensile) stress, also causes an increase (decrease) of the bandgap.

The temperature dependence of the energy bandgap, E_g, has been experimentally determined yielding the following expression for E_g as a function of the temperature, T:

$$E_g(T) = E_g(0) - \frac{\alpha T^2}{T + \beta} \qquad \text{...(1.5)}$$

where E_g (0), α and β are the fitting parameters. These fitting parameters are listed for germanium, silicon and gallium arsenide in Table 1.1:

Table 1.1 Parameters used to calculate the energy bandgap of germanium, silicon and gallium arsenide (GaAs) as a function of temperature

	Germanium	*Silicon*	*GaAs*
E_g (0) (eV)	0.7437	1.166	1.519
α (meV/K)	0.477	0.473	0.541
β (K)	235	636	204

A plot of the resulting bandgap versus temperature is shown in Fig. 1.8 for germanium, silicon and gallium arsenide.

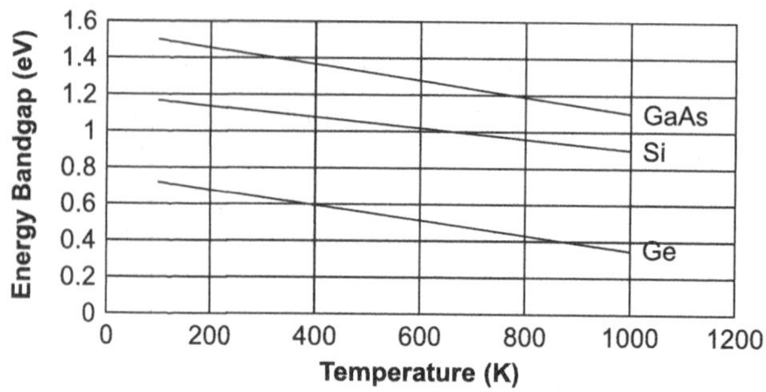

Fig. 1.8 *Temperature dependence of the energy bandgap of germanium (Ge), silicon (Si) and gallium arsenide (GaAs)*

Example: *Calculate the energy bandgap of germanium, silicon and gallium arsenide at 300, 400, 500 and 600 K.*

Solution. The bandgap of silicon at 300 K equals:

$$E_g\,(300\text{ K}) = E_g\,(0\text{ K}) - \frac{\alpha T^2}{T+\beta} = 1.66 - \frac{0.473 \times (300)^2}{300 + 636} = 1.12 \text{ eV}$$

Similarly one finds the energy bandgap for germanium and gallium arsenide, as well as at different temperatures, yielding:

	Germanium	*Silicon*	*Gallium Arsenide*
T = 300 K	0.66 eV	1.12 eV	1.42 eV
T = 400 K	0.62 eV	1.09 eV	1.38 eV
T = 500 K	0.58 eV	1.06 eV	1.33 eV
T = 600 K	0.54 eV	1.03 eV	1.28 eV

1.3.4 Doping Dependence of the Energy Bandgap

High doping densities cause the bandgap to shrink. This effect is explained by the fact that the wavefunctions of the electrons bound to the impurity atoms start to overlap as the density of the impurities increase. For instance, at a doping density of 10^{18} cm^{-3}, the average distance between two impurities is only 10 nm. This overlap forces the energies to form an energy band rather than a discreet level.

If the impurity level is shallow, this impurity band reduces the energy band of the host material by:

$$\Delta E_g\,(N) = -\frac{3q^2}{16\pi\varepsilon_s}\sqrt{\frac{q^2 N}{\varepsilon_s kT}} \qquad \text{... (1.6)}$$

where N is the doping density, q is the electronic charge, ε_s is the dielectric constant of the semiconductor, k is Boltzmann's constant and T is the temperature in Kelvin. For silicon (ε_r = 11.7) this expression further reduces to:

$$\Delta E_g(N) = -22.5\sqrt{\frac{N}{10^{18}\ \text{cm}^{-3}}}\ \text{meV} \qquad \text{... (1.7)}$$

From this expression we find that the bandgap shrinkage can typically be ignored for doping densities less than 10^{18} cm^{-3}.

A plot of the change in bandgap energy with doping density is shown in Fig. 1.9:

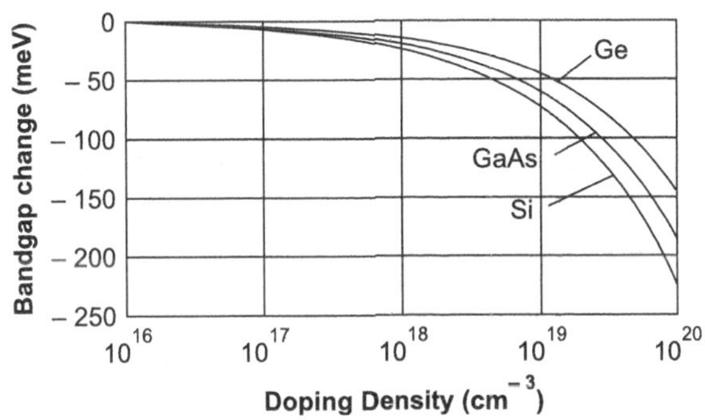

Fig. 1.9 *Doping dependence of the energy bandgap of germanium (Ge), gallium arsenide (GaAs)*

1.4 METALS, INSULATORS AND SEMICONDUCTORS

Once we know the bandstructure of a given material we still need to find out which energy levels are occupied and whether specific bands are empty, partially filled or completely filled.

Empty bands do not contain electrons. Therefore, they are not expected to contribute to the electrical conductivity of the material. Partially filled bands do contain electrons as well as available energy levels at slightly higher energies. These unoccupied energy levels enable carriers to gain energy when moving in an applied electric field. Electrons in a partially filled band therefore do contribute to the electrical conductivity of the material.

Completely filled bands do contain plenty of electrons but do not contribute to the conductivity of the material. This is because the electrons cannot gain energy since all energy levels are already filled.

In order to find the filled and empty bands we must find out how many electrons can be placed in each band and how many electrons are available. Each band is formed due to the splitting of one or more atomic energy levels. Therefore, the minimum number of states in a band equals twice the number of atoms in the material. The reason for the factor of two is that every energy level can contain two electrons with opposite spin.

To further simplify the analysis, we assume that only the *valence* electrons (the electrons in the outer shell) are of interest. The core electrons are tightly bound to the atom and are not allowed to freely move in the material.

Four different possible scenarios are shown in Fig. 1.10:

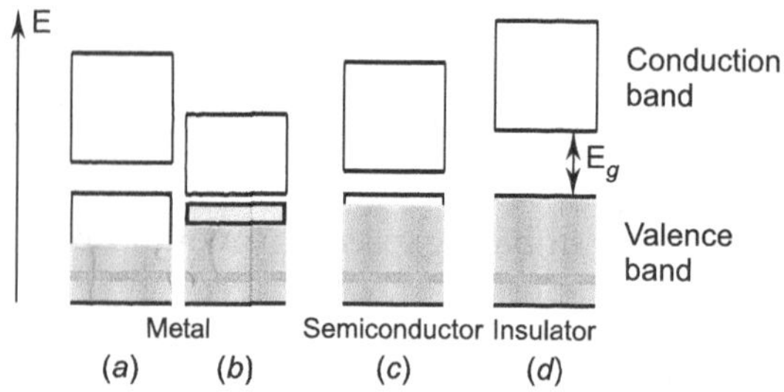

Fig. 1.10 *Possible energy band diagrams of a crystal. Shown are: (a) a half filled band, (b) two overlapping bands, (c) an almost full band separated by a small bandgap from an almost empty band and (d) a full band and an empty band separated by a large bandgap*

A half-filled band is shown in Fig. 1.10 (*a*). This situation occurs in materials consisting of atoms, which contain only one valence electron per atom. Most highly conducting metals including copper, gold and silver satisfy this condition. Materials consisting of atoms that contain two valence electrons can still be highly conducting if the resulting filled band overlaps with an empty band. This scenario is shown in (*b*). No conduction is expected for scenario (*d*) where a completely filled band is separated from the next higher empty band by a larger energy gap. Such materials behave as insulators. Finally, scenario (*c*) depicts the situation in a semiconductor. The completely filled band is now close enough to the next higher empty band that electrons can make it into the next higher band. This yields an almost full band below an almost empty band. We will call the almost full band the *valence band* since it is occupied by valence electrons. The almost empty band will be called the *conduction band*, as electrons are free to move in this band and contribute to the conduction of the material.

1.5 ELECTRONS AND HOLES IN SEMICONDUCTORS

As pointed out in section 1.4, semiconductors differ from metals and insulators by the fact that they contain an "almost-empty" conduction band and an "almost-full" valence band. This also means that we will have to deal with the transport of carriers in both bands.

To facilitate the discussion of the transport in the "almost-full" valence band of a semiconductor, we will introduce the concept of holes. It is important to understand that one should deal with only electrons if one is willing to keep track of all the electrons in the "almost-full" valence band. After all, electrons are the only real particles available in a semiconductor.

The concepts of holes is introduced in semiconductors since it is easier to keep track of the missing electrons in an "almost-full" band, rather than keeping track of the actual electrons in that band. We will now first explain the concept of a hole and then point out how the hole concept simplifies the analysis.

Holes are missing electrons. They behave as particles with the same properties as the electrons would have when occupying the same states except that they carry a positive charge. This definition is illustrated further with Fig. 1.11, which presents the energy band diagram in the presence of an electric field.

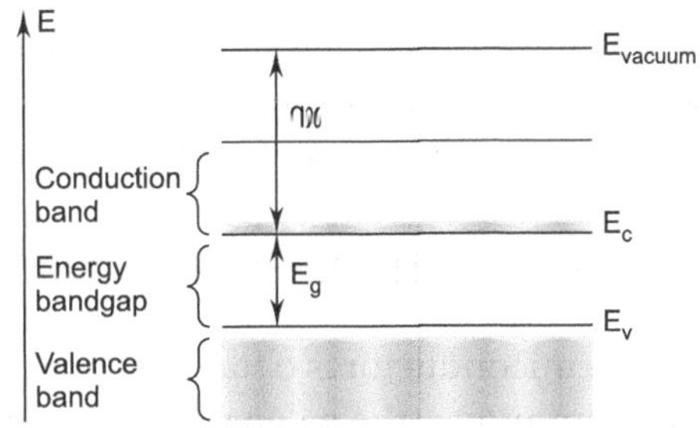

Fig. 1.11 *Energy band diagram in the presence of a uniform electric field. Shown are the upper almost-empty band and the lower almost-filled band. The tilt of the bands is caused by an externally applied electric field*

A uniform electric field is assumed which causes a constant gradient of the bands.

The electrons in the almost-empty band are negatively charged particles, which therefore move in a direction, which opposes the direction of the field. Electrons therefore move *down hill* in the upper band. Electrons in the lower band also move in the same direction. The total current density due to the electrons in the valence band can therefore be written as:

$$J_{vb} = \frac{1}{V} \sum_{\substack{\text{filled} \\ \text{states}}} (-q)\, v_i \quad \text{... (1.8)}$$

where V is the volume of the semiconductor, q is the electronic charge and v is the electron velocity. The sum is taken overall occupied or filled states in the lower band. This equation can be reformulated by first taking the sum overall the states in the lower band and subtracting the current due to the electrons, which are missing in the almost-filled band. This last term therefore represents the sum taken overall the empty states in the lower band,

$$J_{vb} = \frac{1}{V} \left(\sum_{\substack{\text{all} \\ \text{states}}} (-q)\, v_i - \sum_{\substack{\text{empty} \\ \text{states}}} (-q)\, v_i \right) \quad \text{... (1.9)}$$

The sum of all the states in the lower band has to equal zero since electrons in a completely filled band do not contribute to current, while the remaining term can be written as:

$$J_{vb} = \frac{1}{V} \sum_{\substack{\text{empty} \\ \text{states}}} (+q)\, v_i \quad \text{... (1.10)}$$

which states that the current is due to positively charged particles associated with the empty states in the almost-filled band. We call these particles *holes*. Keep in mind that there is no real particle associated with a hole. Instead, the combined behavior of all the electrons, which occupy states in the almost-filled band, is the same as that of positively charged particles associated with the unoccupied states.

The reason the concept of holes simplifies the analysis is that the density of states function of a whole band can be rather complex. However, it can be dramatically simplified if only states close to the band edge need to be considered.

1.6 DETAILED DESCRIPTION OF THE EFFECTIVE MASS CONCEPT

The effective mass of a semiconductor is obtained by fitting the actual E-k diagram around the conduction band minimum or the valence band maximum by a paraboloid. While this concept is simple enough, the issue turns out to be substantially more complex due to the multitude and the occasional anisotropy of the minima and maxima. In this section we first describe the different relevant band minima and maxima, present the numeric values for germanium, silicon and gallium arsenide and introduce the effective mass for density of states calculations and the effective mass for conductivity calculations.

Most semiconductors can be described as having one band minimum at $k = 0$ as well as several equivalent anisotropic band minima at $k \neq 0$. In addition there are three band maxima of interest close to the valence band edge.

1.6.1 Band Structure of Silicon

As an example we consider the band structure of silicon as shown in the Fig. 1.12 below:

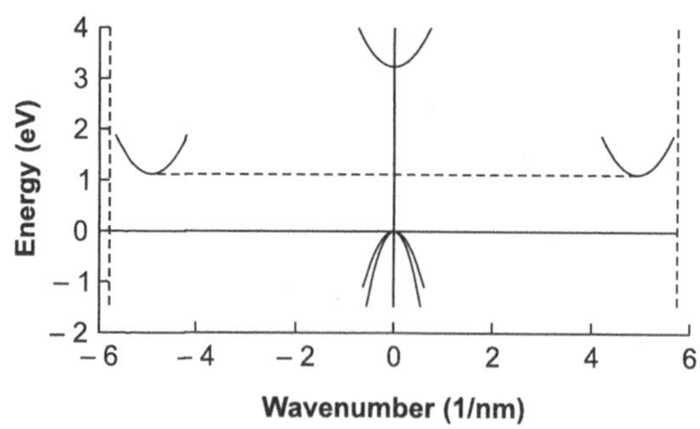

Fig. 1.12 *Simplified E-k diagram of silicon*

Shown is the E-k diagram within the first brillouin zone and along the (100) direction. The energy is chosen to be to zero at the edge of the valence band. The lowest band minimum at $k = 0$ and still above the valence band edge occurs at $E_{c,\,direct} = 3.2$ eV. This is not the lowest minimum above the valence band edge since there are also 6 equivalent minima at $k = (x, 0, 0)$, $(-x, 0, 0)$, $(0, x, 0)$, $(0, -x, 0)$, $(0, 0, x)$, and $(0, 0, -x)$ with $x = 5$ nm^{-1}. The minimum energy of all these minima equals 1.12 $eV = E_{c,\,indirect}$. The effective mass of these anisotropic minima is characterized by a longitudinal mass along the corresponding equivalent (100) direction and two transverse masses in the plane perpendicular to the longitudinal direction. In silicon the longitudinal electron mass is $m_{e,\,l}^* = 0.98\ m_0$ and the transverse electron masses are $m_{e,\,t}^* = 0.19\ m_0$, where $m_0 = 9.11 \times 10^{-31}$ kg is the free electron rest mass.

Two of the three band maxima occur at 0 eV. These bands are referred to as the light and heavy hole bands with a light hole mass of $m_{l,h}^* = 0.16\ m_0$ and heavy hole mass of $m_{h,h}^* = 0.46\ m_0$. In addition there is a split-off hole band with its maximum at $E_{v,so} = -0.044$ eV and a split-off hole mass of $m_{h,so}^* = 0.29\ m_0$.

1.6.2 Effective Mass and Energy Band Minima and Maxima of Ge, Si and GaAs

The values of the energy band minima and maxima as well as the effective masses for germanium, silicon and gallium arsenide are listed in the table below:

Table 1.2 Effective mass of carriers in germanium, silicon and gallium arsenide (GaAs)

Name	*Symbol*	*Germanium*	*Silicon*	*Gallium Arsenide*
Band minimum at k = 0				
Minimum energy	$E_{g,\,direct}$ (eV)	0.8	3.2	1.424
Effective mass	m_e^*/m_0	0.041	0.2	0.067
Band minimum not at k = 0				
Minimum energy	$E_{g,\,indirect}$ (eV)	0.66	1.12	1.734
Longitudinal effective mass	$m_{e,l}^*/m_0$	1.64	0.98	1.98
Transverse effective mass	$m_{e,t}^*/m_0$	0.082	0.19	0.37
Longitudinal direction		– 111	– 100	– 111
Heavy hole valence band maximum at E = k = 0				
Effective mass	m_{hh}^*/m_0	0.28	0.49	0.45
Light hole valence band maximum at k = 0				
Effective mass	m_{lh}^*/m_0	0.044	0.16	0.082
Split-off hole valence band maximum at k = 0				
Split-off hole valence band enegy	$E_{v,\,so}$ (eV)	– 0.028	– 0.044	– 0.34
Effective mass	$m_{h,\,so}^*/m_0$	0.084	0.29	0.154

1.6.3 Effective Mass for Density of States Calculations

The effective mass for density of states calculations equals the mass which provides the density of states using the expression for one isotropic maximum or minimum or:

$$g_c(E) = \frac{8\pi\sqrt{2}}{h^3} m_e^{3/2} \sqrt{E - E_c} \text{ for } E \geq E_c \quad \text{... (1.11)}$$

for the density of states in the conduction band and:

$$g_v(E) = \frac{8\pi\sqrt{2}}{h^3} m_h^{3/2} \sqrt{E_v - E} \text{ for } E \leq E_v \quad \text{... (1.12)}$$

for the density of states in the valence band.

For instance, for a single band minimum described by a longitudinal mass and two transverse masses the effective mass for density of states calculations is the geometric mean of the three masses. Including the fact that there are several equivalent minima at the same energy one obtains the effective mass for density of states calculations from:

$$m^*_{e.dos} = M_c^{3/2}\sqrt[3]{m_1 m_t m_t} \quad \text{... (1.13)}$$

where M_c is the number of equivalent band minima. For silicon one obtains:

$$m^*_{e.dos} = M_c^{3/2}\sqrt[3]{m_1 m_t m_t} = 6^{3/2}\sqrt[3]{0.89\times 0.19\times 0.19}\ m_0 = 1.08\ m_0 \quad \text{... (1.14)}$$

1.6.4 Effective Mass for Conductivity Calculations

The effective mass for conductivity calculation is the mass, which is used in condition related problems accounting for the detailed structure of the semiconductor. These calculations include mobility and diffusion constants calculations. Another example is the calculation of the shallow impurity levels using a hydrogen-like model.

As the conductivity of a material is inversionally proportional to the effective masses, one finds that the conductivity due to multiple band maxima or minima is proportional to the sum of the inverse of the individual masses, multiplied by the density of carriers in each band, as each maximum or minimum adds to the overall conductivity. For anisotropic minima containing one longitudinal and two transverse effective masses one has to sum over the effective masses in the different minima along the equivalent directions. The resulting effective mass for bands, which have ellipsoidal constant energy surfaces, is given by:

$$m^*_{e.cond} = \frac{3}{\dfrac{1}{m_1}+\dfrac{1}{m_t}+\dfrac{1}{m_t}} \quad \text{... (1.15)}$$

provided the material has an isotropic conductivity as is the case for cubic materials. For instance electrons in the X minima of silicon have an effective conductivity mass given by:

$$m^*_{e.cond} = \frac{3}{\dfrac{1}{m_1}+\dfrac{1}{m_t}+\dfrac{1}{m_t}} = \frac{3}{\dfrac{1}{0.89}+\dfrac{1}{0.19}+\dfrac{1}{0.19}} = 0.26\ m_0 \quad \text{... (1.16)}$$

1.7 DENSITY OF STATES

Before we can calculate the density of carriers in a semiconductor, we have to find the number of available states at each energy. The number of electrons at each energy is then obtained by multiplying the number of states with the probability that a state is occupied by an electron. Since the number of energy levels is very large and dependent on the size of the semiconductor, we will calculate the number of states per unit energy and per unit volume.

1.7.1 Calculation of the Density of States

The density of states in a semiconductor equals the density per unit volume and energy of the number of solutions to Schrodinger's equation. We will assume that the semiconductor can be modeled as an infinite quantum well in which electrons with effective mass, m*, are free to move. The energy in he well is set to zero. The semiconductor is assumed a cube with side L. This assumption does not affect the result since the density of states per unit volume should not depend on the actual size or shape of the semiconductor.

The solutions to the wave equation where V $(x) = 0$ are sine and cosine functions:

$$\Psi = A \sin (k_x x) + B \cos (k_x x) \quad ...(1.17)$$

Where A and B are to be determined. The wavefunction must be zero at the infinite barriers of the well. At $x = 0$ the wavefunction must be zero so that only sine functions can be valid solutions or B must equal zero. At $x = L$, the wavefunction must also be zero yielding the following possible values for the wavenumber, k_x.

$$k_x = \frac{n\pi}{L} . n = 1, 2, 3, ... \quad ...(1.18)$$

This analysis can now be repeated in the y and z direction. Each possible solution then corresponds to a cube in k-space with size $\frac{n\pi}{L}$ as indicated on Fig. 1.13.

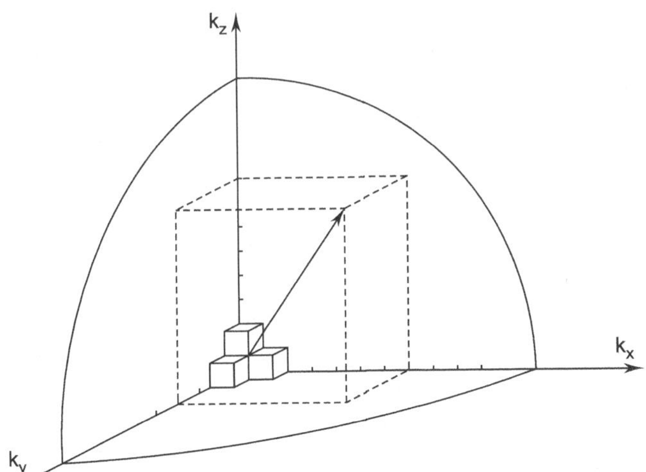

Fig. 1.13 *Calculation of the number of states with wavenumber less than k*

The total number of solutions with a different value for k_x, k_y and k_z and with a magnitude of the wavevector less than k is obtained by calculating the volume of one eighth of a sphere with radius k and dividing it by the volume corresponding to a single solution, $\left(\frac{\pi}{L}\right)^3$, yielding:

$$N = 2 \times \frac{1}{8} \times \left(\frac{L}{\pi}\right)^3 \times \frac{4}{3} \times \pi \times k^3 \quad ...(1.19)$$

A factor of two is added to account for the two possible spins of each solution. The density per unit energy is then obtained using the chain rule:

$$\frac{dN}{dE} = \frac{dN}{dk}\frac{dk}{dE} = \left(\frac{L}{\pi}\right)^3 \pi k^2 \frac{dk}{dE} \quad \text{... (1.20)}$$

The kinetic energy E of a particle with mass m^* is related to the wavenumber, k, by:

$$E(k) = \frac{\hbar^2 k^2}{2m^*}, \text{providing } \frac{dK}{dE} = \frac{m^*}{\hbar^2 k} \text{ and } k = \frac{\sqrt{2m^* E}}{\hbar} \quad \text{... (1.21)}$$

And the density of states per unit volume and per unit energy, g (E), becomes:

$$g(E) = \frac{1}{L^3}\frac{dN}{dE} = \frac{8\pi\sqrt{2}}{h^3} m^{*3/2} \sqrt{E} \text{ for } E \geq 0 \quad \text{... (1.22)}$$

The density of states is zero at the bottom of the well as well as for negative energies.

The same analysis also applies to electrons in a semiconductor. The effective mass takes into account the effect of the periodic potential on the electron. The minimum energy of the electron is the energy at the bottom of the conduction band, E_c, so that the density of states for electrons in the conduction band is given by:

$$g_c(E) = \frac{8\pi\sqrt{2}}{h^3} m^{*3/2} \sqrt{E - E_c} \text{ for } E \geq E_c \quad \text{... (1.23)}$$

$$g_c(E) = 0, \text{ for } E < E_c$$

Example. *Calculate the number of states per unit energy in a 100 by 100 by 10 nm piece of silicon ($m^* = 1.08\ m_0$) 100 meV above the conduction band edge. Write the result in units of eV^{-1}.*

Solution. The density of states equals:

$$g(E) = \frac{8\pi\sqrt{2}}{h^3} m^{*3/2} \sqrt{E - E_c}$$

$$= \frac{8\pi\sqrt{2}\left(1.08 \times 9 \times 10^{-31}\right)^{3/2}}{\left(6.626 \times 10^{-34}\right)^3} \sqrt{0.1 \times 1.6 \times 10^{-19}}$$

$$= 1.51 \times 10^{56}\ m^{-3}\ J^{-1}.$$

So that the total number of states per unit energy equals:

$$g(E)\,V = 1.51 \times 10^{56} \times 10^{-22}\ J^{-1} = 2.41 \times 10^5\ eV^{-1}$$

1.7.2 Calculation of the Density of States in 1, 2 and 3 Dimensions

We will here postulate that the density of electrons in k-space is constant and equals the physical length of the sample divided by 2π and that for each dimension. The number of states between k and $k + dk$ in 3, 2 and 1 dimension then equals:

$$\frac{dN_{3D}}{dk} = 2\left(\frac{L}{2\pi}\right)^3 4\pi k^2 \,.\, \frac{dN_{2D}}{dk} = 2\left(\frac{L}{2\pi}\right)^2 . 2\pi k \frac{dN_{1D}}{dk} = 2\left(\frac{L}{2\pi}\right) \quad \text{... (1.24)}$$

We now assume that the electrons in a semiconductor are close to a band minimum, E_{min} and can be described as free particles with a constant effective mass, or:

$$E(k) = E_{min} + \frac{h^2 k^2}{8\pi^2 m^*} \quad \text{... (1.25)}$$

Elimination of k using the E (k) relation above then yields the desired density of states function, namely:

$$g_{c,3D} = \frac{dN_{3D}}{dE} = \frac{8\pi\sqrt{2}}{h^3} m^{*3/2} \sqrt{E - E_{min}}, \text{ for } E \geq E_{min} \quad \text{... (1.26)}$$

for a three-dimensional semiconductor,

$$g_{c,2D} = \frac{dN_{2D}}{dE} = \frac{4\pi m^*}{h^2}, \text{ for } E \geq E_{min} \quad \text{... (1.27)}$$

For a two-dimensional semiconductor such as a quantum wire in which particles are confined to a plane, and gc, 1D = $\frac{dN_1 D}{dE} = \sqrt{\frac{2\pi m^*}{h^2}} \frac{1}{\sqrt{E - E_{min}}}$ when $E \geq E_{min}$ ... (1.28)

For a one-dimensional semiconductor such as a quantum wire in which particles are confined along a line. An example of the density of states in 3, 2 and 1 dimension is shown in the Fig. 1.14 below:

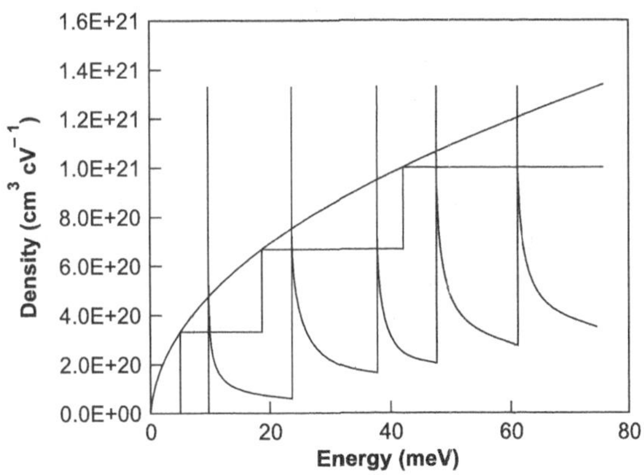

Fig. 1.14 *Density of states per unit volume and energy for a 3-D semiconductor (blue curve), a 10 nm quantum well with infinite barriers (red curve) and a 10 nm by 10 nm quantum wire with infinite barriers (green curve), $m^*/m_0 = 0.8$*

The above Fig. 1.14 illustrates the added complexity of the quantum well and quantum wire: Even though the density in two dimensions is constant, the density of states for a quantum well is a step function with steps occurring at the energy of each quantized level. The case for the quantum wire is further complicated by the degeneracy of the energy levels: For instance a two-fold degeneracy increases the density of states associated with that energy level by a factor of two. A list of the degeneracy (not including spin) for the 10 lowest energies in a quantum well, a quantum wire and a quantum box, all with infinite barriers, is provided in the table as follows:

Table 1.3 Degeneracy (not including spin) of the lowest 10 energy levels in a quantum well, a quantum wire with square cross-section and a quantum cube with infinite barriers. The energy E_0 equals the lowest energy in a quantum well, which has the same size

Quantum Well			*Quantum Wire*				*Quantum Box*				
n_x	E/E_0	*Degeneracy*	n_x	n_y	E/E_0	*Degeneracy*	n_x	n_y	n_z	E/E_0	*Degeneracy*
1	1	1	1	1	2	1	1	1	1	3	1
2	4	1	1	2	5	2	1	1	2	6	3
3	9	1	2	2	8	1	1	2	2	9	3
4	16	1	1	3	10	2	1	1	3	11	3
5	25	1	2	3	13	2	2	2	2	12	1
6	36	1	1	4	17	2	1	2	3	14	6
7	49	1	3	3	18	1	2	2	3	17	3
8	64	1	2	4	20	2	1	1	4	18	3
9	81	1	3	4	25	2	1	3	3	19	3
10	100	1	1	5	26	2	1	2	4	21	6
11	121	1	2	5	29	2	2	3	3	22	3
12	144	1	4	4	32	1	2	2	4	24	6
13	169	1	3	5	34	2	1	3	4	26	3
14	196	1	1	6	37	6	1	1	5	27	3
15	225	1	2	6	40	13	3	3	3	27	1
							3	2	4	29	3

Next, we compare the actual density of states in three dimensions with equation (1.26). While somewhat tedious, the exact number of states can be calculated as well as the maximum energy. The result is shown in Fig. 1.15. The number of states in an energy range of 20 E_0 are plotted as a function of the normalized energy E/E_0. A dotted line is added to guide the eye. The solid line is calculated using equation (1.26). A clear difference can be observed between the two, while they are expected to merge for large values of E/E_0.

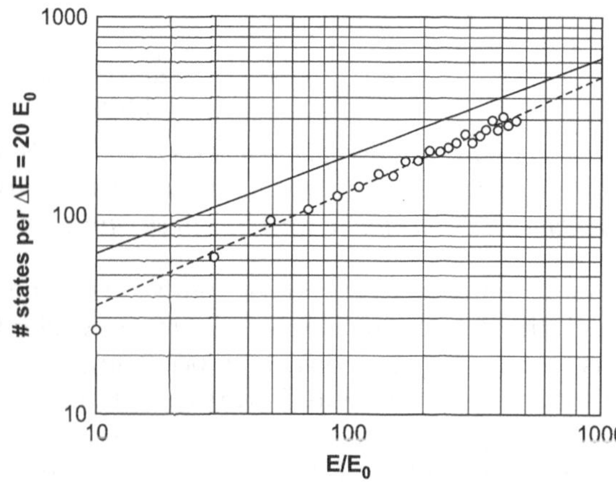

Fig. 1.15 *Number of states within a range $\Delta E = 20\, E_0$ as a function of the normalized energy E/E_0. (E_0 is the lowest energy in a 1-dimensional quantum well). See text for more detail.*

A comparison of the total number of states illustrates the same trend as shown in Fig. 1.16. Here the solid line indicates the actual number of states, while the dotted line is obtained by integrating equation (1.26).

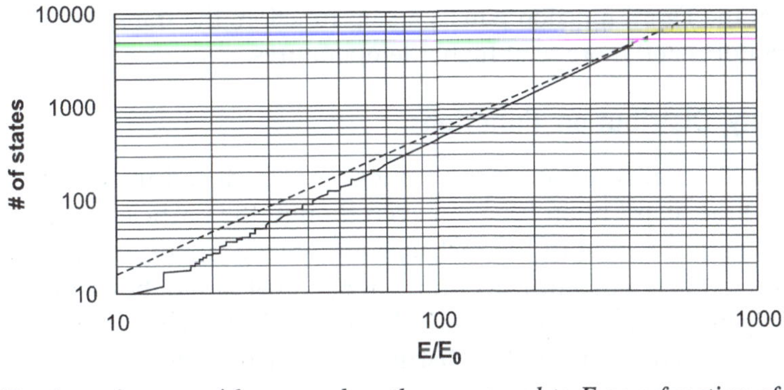

Fig. 1.16 *Number of states with energy less than or equal to E as a function of E_0 (E_0 is the lowest energy in an 1-dimensional quantum well). Actual number (solid line) is compared with the integral of equation (1.26) (dotted line).*

1.8 CARRIER DISTRIBUTION FUNCTIONS

The distribution or probability density functions describe the probability that particles occupy the available energy levels in a given system. Of particular interest if the probability density function of electrons, called the Fermi function. The derivation of such probability density functions can be found in one of the many statistical thermodynamics references. However, given the importance of the Fermi distribution function, we will carefully examine an example as well as the characteristics of this function. It is also derived in section 1.8.5. Other distribution functions such as the impurity distribution functions, the Bose-Einstein distribution function and the Maxwell Boltzmann distribution are also provided.

1.8.1 Fermi-Dirac Distribution Function

The Fermi-Dirac distribution function, also called Fermi function, provides the probability of occupancy of energy levels by Fermions. Fermions are half-integer spin particles, which obey the Pauli exclusion principle. The Pauli exclusion principle postulates that only one Fermion can occupy a single quantum state. Therefore, as Fermions are added to an energy band, they will fill the available states in an energy band just like water fills a bucket. The states with the. lowest energy are filled first, followed by the next higher ones. At absolute zero temperature (T = 0 K), the energy levels are all filled up to a maximum energy, which we call the Fermi level. No states above the Fermi level are filled. At higher temperature, one finds that the transition between completely filled states and completely empty states is gradual rather than abrupt.

$$f(E) = \frac{1}{1 + e^{(E - E_F)/kT}} \qquad \text{... (1.29)}$$

The Fermi function has a value of one for energies, which are more than a few times kT below the Fermi energy. It equals 1/2 if the energy equals the Fermi energy and decreases exponentially for energies which are a few times kT larger than the Fermi energy. While at T = 0 K the Fermi function equals a step function, the transition is more gradual at finite temperatures and more so at higher temperatures.

1.8.2 Impurity Distribution Functions

The distribution function of impurities, as introduced in section 1.9.4 differs from the Fermi-Dirac distribution function even though the particles involved are still Fermions. The difference is due to the fact that an ionized donor energy level still contains one electron with either spin. The donor energy level cannot be empty since this would leave a doubly positively charged atom, which would have an energy different from that of the singly ionized donor level. The distribution function for donors therefore differs from the Fermi function and is given by:

$$f_{donor}(E_d) = \frac{1}{1+\frac{1}{2}e^{(E_d - E_F)/kT}} \quad \text{... (1.30)}$$

The distribution function for acceptors differs also because of the different possible ways to occupy the acceptor level. The neutral acceptor contains no electrons. The ionized acceptor contains one electron, which can have either spin, while the doubly negatively charged state is not allowed since this would require a different energy. This restriction would yield a factor of 2 in front of the exponential term. In addition, one finds that most commonly used semiconductors have a two-fold degenerate valence band, which causes this factor to increase to four, yielding:

$$f_{acceptor}(E_a) = \frac{1}{1+4e^{(E_a - E_F)/kT}} \quad \text{... (1.31)}$$

1.8.3 Other Distribution Functions and Comparison

Other distribution functions include the Bose-Einstein distribution and the Maxwell-Boltzmann distribution. These are briefly discussed below and compared to the Fermi-Dirac distribution function.

The Bose-Einstein distribution function applies to bosons. Bosons are particles with integer spin and include photons, phonons and a large number of atoms. Bosons do not obey the Pauli exclusion principle so that any number can occupy a single energy level. The Bose-Einstein distribution function is given by:

$$f_{BE}(E) = \frac{1}{e^{(E - E_F)/kT - 1}} \quad \text{... (1.32)}$$

This function is only defined for $E > E_F$.

The Maxwell Boltzmann applies to non-interacting particles, which can be distinguished from each other. This distribution function is also called the classical distribution since it provides the probability of occupancy for non-interacting particles at low densities. Atoms in an ideal gas form a typical example of such particles. The Maxwell-Boltzmann distribution function is given by:

$$f_{MB}(E) = \frac{1}{e^{(E-E_f)/kT}} \quad \text{... (1.33)}$$

A plot of the three distribution functions, the Fermi-Dirac distribution, the Maxwell-Boltzmann distribution and the Bose-Einstein distribution is shown in Fig. 1.17.

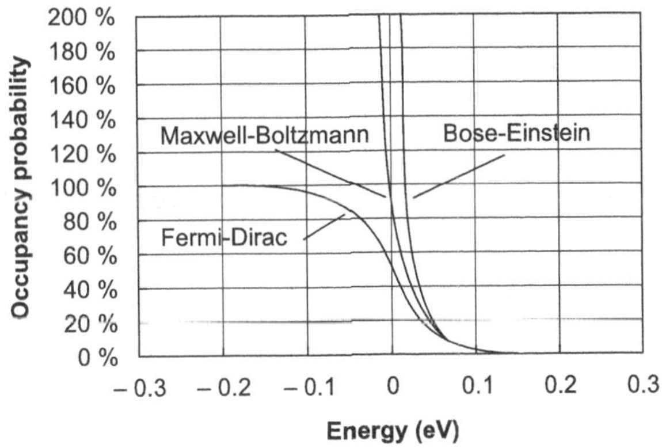

Fig. 1.17 *Probability of occupancy versus energy of the Fermi-Dirac, the Bose-Einstein and the Maxwell-Boltzmann distribution. The Fermi energy, E_F, is assumed to be zero.*

All three functions are almost equal for large energies (more than a few kT beyond the Fermi energy). The Fermi-Dirac distribution reaches a maximum of 100% for energies, which are a few kT below the Fermi energy, while the Bose-Einstein distribution diverges at the Fermi energy and has no validity for energies below the Fermi energy.

1.8.4 Derivation of the Fermi-Dirac Distribution Function

To derive the Fermi-Dirac distribution function, we start from a series of possible energies, labeled E_i. At each energy, we can have g_i possible states and the number of states that are occupied equals $g_i f_i$, where f_i is the probability of occupying a state at energy E_i. We also assume that the number of possible states is very large, so that the discrete nature of the states can be ignored.

The number of possible ways—called configurations—to fit $g_i f_i$ electrons in g_i states, given the restriction that only one electron can occupy each state, equals:

$$W_i = \frac{g_i!}{(g_i - g_i f_i)!\, g_i f_i!} \quad \text{... (1.34)}$$

This equation is obtained by numbering the individual states and exchanging the states rather than the electrons. This yields a total number of $g_i!$ possible configurations. However, since the empty states are all identical, we need to divide by the number of permutations between the empty states, as all permutations cannot be distinguished from each other and can therefore only be counted once. In addition, all the filled states are indistinguishable from each other, so we need to divide also by all permutations between the filled states, namely $g_i f_i!$

The number of possible ways to fit the electrons in the number of available states is called the multiplicity function.

The multiplicity function for the whole system is the product of the multiplicity functions for each energy E_i.

$$W = \prod_i W_i = \prod_i \frac{g_i!}{(g_i - g_i f_i)!\, g_i f_i!} \qquad \text{... (1.35)}$$

Using Stirling's approximation, one can eliminate the factorial signs, yielding:

$$\ln W = \sum_i \ln W_i = \sum_i \left[g_i \ln g_i - g_i (1 - f_i) \ln (g_i - g_i f_i) - g_i f_i \ln g_i f_i \right] \qquad \text{... (1.36)}$$

The total number of electrons in the system equals N and the total energy of those N electrons equals U. These system parameters are related to the number of states at each energy, g_i, and the probability of occupancy of each state, f_i, by:

$$N = \sum_i g_i f_i \qquad \text{... (1.37)}$$

and

$$U = \sum_i E_i g_i f_i \qquad \text{... (1.38)}$$

According to the basic assumption of statistical thermodynamics, all possible configurations are equally probable. The multiplicity function provides the number of configurations for a specific set of occupancy probabilities, f_i. The multiplicity function sharply peaks at the thermal equilibrium distribution since this is the most likely distribution of the system and must therefore be associated with the largest number of equally probable configurations. The occupancy probability in thermal equilibrium is therefore obtained by finding the maximum of the multiplicity function, W, while keeping the total energy and the number of electrons constant.

For convenience, we maximize the logarithm of the multiplicity function instead of the multiplicity function itself. According to the Lagrange method of undetermined multipliers, we must maximize the following function:

$$\ln W - a \sum_j g_j f_j - b \sum_j E_j g_j f_j \qquad \text{... (1.39)}$$

where a and b need to be determined. The maximum of the multiplicity function, W, is obtained from:

$$\frac{\partial}{\partial (g_i f_i)} \left[\ln W - a \sum_j g_j f_j - b \sum_j E_j g_j f_j \right] = 0 \qquad \text{... (1.40)}$$

which can be solved, yielding:

$$\ln \frac{g_i - g_i f_i}{g_i f_i} - a - bE_i = 0 \qquad \text{... (1.41)}$$

or

$$f_i = f_{FD}(E_i) = \frac{1}{1 + \exp(a + bE_i)} \qquad \text{... (1.42)}$$

which can be written in the following form

$$f_{FD}(E_i) = \frac{1}{1 + \exp\left(\frac{E_i - E_F}{\beta}\right)} \qquad \text{... (1.43)}$$

with $= \frac{1}{b}$ and $E_F = -\frac{a}{b}$. The symbol E_F was chosen since this constant has units of energy and will be the constant associated with this probability distribution.

Taking the derivative of the total energy, one obtains:

$$dU = \sum_i E_i \, d(g_i f_i) + \sum_i g_i f_i \, dE_i \qquad \text{... (1.44)}$$

Using the Lagrange equation, this can be rewritten as:

$$dU = \beta d\,(\ln W) + \sum_i g_i f_i \, dE_i + E_F \, dN \qquad \text{... (1.45)}$$

Any variation of the energies, E_1, can only be caused by a change in volume, so that the middle term can be linked to a volume variation dV.

$$dU = \beta d\,(\ln W) + \left[\sum_i g_i f_i \frac{dE_i}{dV}\right] dV + E_F \, dN \qquad \text{... (1.46)}$$

Comparing this to the thermodynamic identity:

$$dU = TdS - pdV + \mu dN \qquad \text{... (1.47)}$$

one finds that $k_t = kT$ and $S = k \ln W$, where k is a constant that must be determined. The energy, E_F, equals the energy associated with the particles, namely the electro-chemical potential, k_t.

The comparison also identifies the entropy, S, as being proportional to the logarithm of the multiplicity function, W. The proportionality constant, k, is known as Boltzmann's constant.

The Fermi-Dirac distribution function then becomes:

$$f_{FD}(E) = \frac{1}{1 + \exp\left(\frac{E - E_F}{kt}\right)} \qquad \text{... (1.48)}$$

Note that this derivation can only truly be followed if one has prior knowledge of statistical thermodynamics. Those who are well versed in this field can quickly derive the Fermi-Dirac and other distribution functions using the Gibbs sum. A more complete description can be found in: Thermal Physics, second edition, C. Kittel and H. Kroemer, Freeman, 1980.

1.9 CARRIER DENSITIES

Now that we have discussed the density of states and the distribution function, we have all the necessary tools to calculate the carrier density in a semiconductor.

1.9.1 General Discussion

The density of electrons in a semiconductor is related to the density of available states and the probability that each of these states is occupied. The density of occupied states per unit volume and energy, n (E), is simply the product of the density of states in the conduction band, g_c (E) and the Fermi-Dirac probability function, f (E), (also called the Fermi function):

$$n(\mathrm{E}) = g_c(\mathrm{E}) f(\mathrm{E}) \qquad \ldots (1.49)$$

Since holes correspond to empty states in the valence band, the probability of having a hole equals the probability that a particular state is not filled, so that the hole density per unit energy, p (E), equals:

$$p(\mathrm{E}) = g_v(\mathrm{E})\,[1 - f(\mathrm{E})] \qquad \ldots (1.50)$$

Where g_v (E) is the density of states in the valence band. The density of carriers is then obtained by integrating the density of carriers per unit energy overall possible energies within a band. A general expression is derived as well as an approximate analytic solution, which is valid for non-degenerate semiconductors. In addition, we also present the Joyce-Dixon approximation, an approximate solution useful when describing degenerate semiconductors.

The density of states in a semiconductor was obtained by solving the Schrodinger equation for the particles in the semiconductor. Rather than using the actual and very complex potential in the semiconductor, we use the simple particle-in-a box model, where one assumes that the particle is free to move within the material.

For an electron which behaves as a free particle with effective mass, m^*, the density of states was derived in section 1.7 yielding:

$$g_c(\mathrm{E}) = \frac{8\pi\sqrt{2}}{h^3} m_e^{*3/2} \sqrt{\mathrm{E} - \mathrm{E}_c}\,, \text{ for } \mathrm{E} \geq \mathrm{E}_c \qquad \ldots (1.51)$$

where E_c is the bottom of the conduction band below which the density of states is zero. The density of states for holes in the valence band is given by:

$$g_v(\mathrm{E}) = \frac{8\pi\sqrt{2}}{h^3} m_h^{*3/2} \sqrt{\mathrm{E}_v - \mathrm{E}}\,, \mathrm{E} \leq \mathrm{E}_v \qquad \ldots (1.52)$$

1.9.2 Calculation of the Fermi Integral

The carrier density in a semiconductor, is obtained by integrating the product of the density of states and the probability density function overall possible states. For electrons in the conduction band the integral is taken from the bottom of the conduction band, labeled, E_c, to the top of the conduction band:

$$n = \int_{\mathrm{E}_c}^{\text{top of the conduction band}} n(\mathrm{E})\, d\mathrm{E} = \int_{\mathrm{E}_c}^{\text{top of the conduction band}} g_c(\mathrm{E}) f(\mathrm{E})\, d\mathrm{E} \qquad \ldots (1.53)$$

Where g_c (E) is the density of states in the conduction band and f (E) is the Fermi function.

This general expression is illustrated with Fig. 1.18 for a parabolic density of states function with $E_c = 0$. The figure shows the density of states function, g_c (E), the Fermi function, f (E), as well as the product of both, which is the density of electrons per unit volume and per unit energy, n (E). The integral corresponds to the crosshatched area.

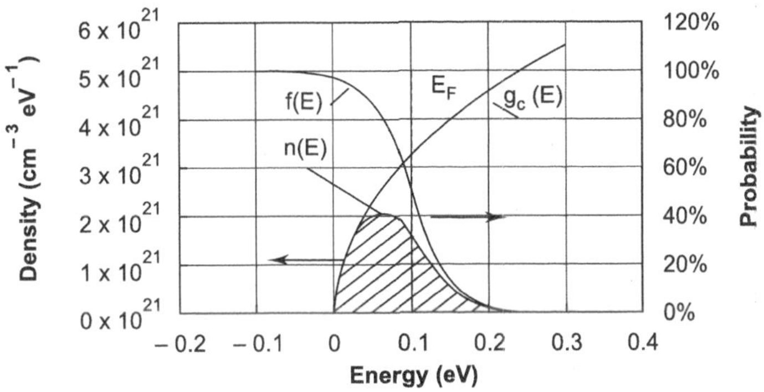

Fig. 1.18 *The carrier density integral. Shown are the density of states, g_c (E), the density per unit energy, n (E) and the probability of occupancy, f (E). The carrier density, n_0, equals the crosshatched area.*

The actual location of the top of the conduction band does not need to be known as the Fermi function goes to zero at higher energies. The upper limit can therefore be replaced by infinity. We also relabeled the carrier density as n_0 to indicate that the carrier density is the carrier density in thermal equilibrium.

$$n_0 = \int_{E_c}^{\infty} g_c(E) f(E)\, dE \qquad \text{... (1.54)}$$

Using equations (1.29) and (1.51) this integral becomes:

$$n_0 = \int_{E_v}^{\infty} \frac{8\pi\sqrt{2}}{h^3} m_e^{*3/2} \sqrt{E_v - E_c}\, \frac{1}{1 + e^{\frac{E - E_f}{kT}}}\, dE \qquad \text{... (1.55)}$$

While this integral cannot be solved analytically at non-zero temperatures, we can obtain either a numeric solution or an approximate analytical solution. Similarly for holes one obtains:

$$p_0 = \int_{-\infty}^{E_v} g_v(E)[1 - f(E)]\, dE \qquad \text{... (1.56)}$$

and

$$p_0 = \int_{-\infty}^{E_v} \frac{8\pi\sqrt{2}}{h^3} m_h^{*3/2} \sqrt{E_v - E}\, \frac{1}{1 + e^{\frac{E_F - E}{kT}}}\, dE \qquad \text{... (1.57)}$$

The calculation of the electron and hole density in a semiconductor is further illustrated by Fig. 1.19.

Indicated in the (Fig. 1.19) are the density of states in the conduction and valence band, the Fermi distribution function and the electron and hole densities per unit energy. The crosshatched areas indicate the thermal-equilibrium carrier densities. From the figure, one can easily see that the electron density will increase as the Fermi energy is increased. The hole density decreases with increasing Fermi energy. As the Fermi energy is decreased, the electron density decreases and the hole density increases.

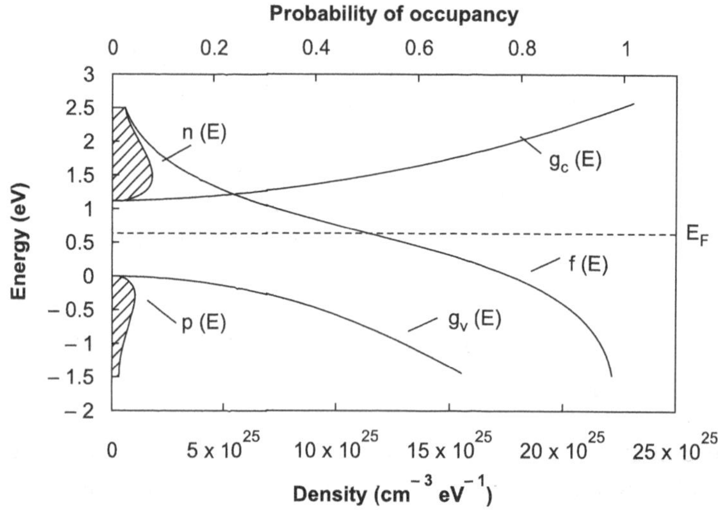

Fig. 1.19 *The density of states and carrier densities in the conduction and valence band. Shown are the electron and hole density per unit energy, n (E) and p (E), the density of states in the conduction and valence band, g_c (E) and g_v (E) and the probability of occupancy, f (E). The crosshatched area indicates the electron and hole densities*

1.9.2.1 Carrier density at zero Kelvin

Equation (1.55) can be solved analytically at T = 0 K, since the Fermi function at T = 0 K equals one for all energies below the Fermi energy and 0 for all energies larger than the Fermi energy. Equation (1.55) can therefore be simplified to:

$$n_0 = \int_{E_c}^{E_F} g_c \,(E)\, dE \text{ at } T = 0\,K \quad \text{... (1.58)}$$

and integration yields:

$$n_0 = \frac{2}{3}\frac{\sqrt{2}}{\pi}\frac{m_c^{*3/2}}{\hbar^3}(E_F - E_c)^{3/2}, \text{ for } E_F \geq E_c \quad \text{... (1.59)}$$

This expression can be used to approximate the carrier density in heavily degenerate semiconductors provided

$$kT << (E_F - E_c) > 0$$

Similarly, on obtains for holes:

$$p_0 = \frac{2}{3}\frac{\sqrt{2}}{\pi^2}\frac{m_h^{*3/2}}{\hbar^2}(E_v - E_F)^{3/2} \text{ for } E_v \geq E_F \quad \text{... (1.60)}$$

1.9.2.2 Non-degenerate semiconductors

Non-degenerate semiconductors are defined as semiconductors for which the Fermi energy is at least 3*k*T away from either band edge. The reason we restrict ourselves to non-degenerate semiconductors is that this definition allows the Fermi function to be replaced with a simple exponential function, *i.e.*, the Maxwell-Boltzmann distribution function. The carrier density integral can then be solved analytically yielding:

$$n_0 \cong \int_{E_c}^{\infty} \frac{8\pi\sqrt{2}}{h^3} m_e^{*3/2} \sqrt{E - E_c}\, e^{\frac{E_F - E}{kT}}\, dE = N_c e^{\frac{E_F - E_c}{KJ}} \quad \text{... (1.61)}$$

with

$$N_c = 2\left[\frac{2\pi m_e^* kT}{h^2}\right]^{3/2} \quad \text{... (1.62)}$$

where N_c is the *effective density of states in the conduction band.* The Fermi energy, E_F, is obtained from:

$$E_F = E_c + kT \ln \frac{n_0}{N_c} \quad \text{... (1.63)}$$

Similarly for holes, one can approximate the hole density integral as:

$$p_0 \cong \int_{-\infty}^{E_v} \frac{8\pi\sqrt{2}}{h^3} m_h^{*3/2} \sqrt{E_v - E}\, e^{\frac{E - E_f}{kT}}\, dE = N_v e^{\frac{E_v - E_f}{kT}} \quad \text{... (1.64)}$$

with

$$N_v = 2\left[\frac{2\pi m_h^* kT}{h^2}\right]^{3/2} \quad \text{... (1.65)}$$

where N_v is the effective density of states in the *valence* band. The Fermi energy, E_F, is obtained from:

$$E_F = E_v - kT \ln \frac{p_0}{N_c} \quad \text{... (1.66)}$$

Example. *Calculate the effective densities of states in the conduction and valence bands of germanium, silicon and gallium arsenide at 300 K.*

Solution. The effective density of states in the conduction band of germanium equals:

$$N_c = 2\left(\frac{2\pi m_e^* kT}{h^2}\right)^{3/2}$$

$$= 2\left(\frac{2\pi \times 0.55 \times 9.11 \times 10^{-31} \times 1.38 \times 10^{-23} \times 300}{\left(6.626 \times 10^{-34}\right)^2}\right)^{3/2}$$

$$= 1.02 \times 10^{25}\ m^{-3} = 1.02 \times 10^{19}\ \text{cm}^{-3}$$

where the effective mass for density of states was used (Appendix 3). Similarly one finds the effective densities for silicon and gallium arsenide and those of the valence band:

	Germanium	Silicon	Gallium Arsenide
N_c (cm^{-3})	1.02×10^{19}	2.81×10^{19}	4.35×10^{17}
N_v (cm^{-3})	5.64×10^{18}	1.83×10^{19}	7.57×10^{18}

Note that the effective density of states is temperature dependent and can be obtained from:

$$N_c(T) = N_c(300\ K)\left(\frac{T}{300}\right)^{3/2}$$

where N_c (300 K) is the effective density of states at 300 K.

1.9.2.3 Degenerate semiconductors

A useful approximate expression applicable to degenerate semiconductors was obtained by Joyce and Dixon and is given by:

$$\frac{E_F - E_c}{kT} = \ln\frac{n_0}{N_c} + \frac{1}{\sqrt{8}}\frac{n_0}{N_c} - \left(\frac{3}{16} - \frac{\sqrt{3}}{9}\right)\left(\frac{n_0}{N_c}\right)^2 + \ldots\ldots \quad \ldots(1.67)$$

for electrons and by:

$$\frac{E_v - E_F}{kT} \cong \ln\frac{p_0}{N_v} + \frac{1}{\sqrt{8}}\frac{p_0}{N_v} - \left(\frac{3}{16} - \frac{\sqrt{3}}{9}\right)\left(\frac{p_0}{N_v}\right)^2 + \ldots\ldots \quad \ldots(1.68)$$

for holes.

A comparison of the Joyce-Dixon approximation is presented in Fig. 1.20.

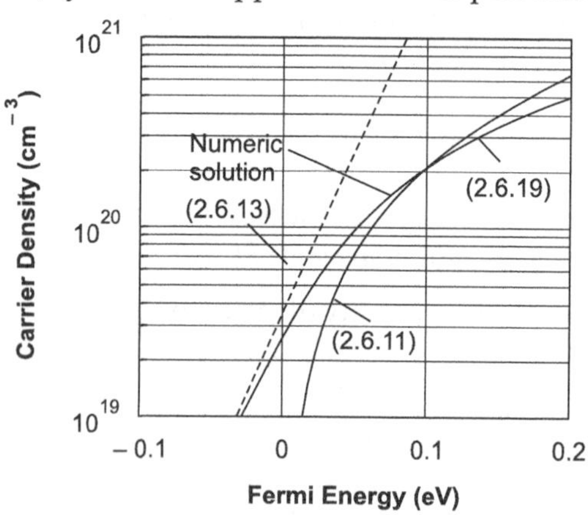

Fig. 1.20 *Carrier (electron) density versus Fermi energy, for $E_c = 0$ eV. Comparison of the Joyce Dixon approximation for degenerate semiconductors with equation (1.51) derived for non-degenerate semiconductors, equation (1.59) derived for $T = 0$ K and a numeric solution*

Compared are the Joyce Dixon approximation with the solution derived for non-degenerate semiconductors, equation (1.61), the solution to the Fermi integral at zero Kelvin, equation (1.59), and a numeric solution. The Joyce Dixon approximation closely follows the numeric solution for all Fermi energies up to 0.1 eV beyond the conduction band edge, and carrier densities above 10^{20} cm^{-3}. The zero temperature solution is a somewhat better fit at higher energies and carrier densities.

1.9.3 Intrinsic Semiconductors

1.9.3.1 Intrinsic carrier density

Intrinsic semiconductors are usually non-degenerate, so that the expressions for the electron (1.60) and hole (1.62) densities in non-degenerate semiconductors apply. Labeling the Fermi energy of intrinsic material as E_i, we can then write two relations between the intrinsic carrier density and the intrinsic Fermi energy, namely:

$$n_i = n_o \big|_{(E_F = E_i)} = N_c e^{(E_i - E_c)/kT}$$

$$n_i = p_o \big|_{(E_F = E_i)} = N_v e^{(E_v - E_i)/kT} \quad \text{... (1.69)}$$

It is possible to eliminate the intrinsic Fermi energy from both equations, simply by multiplying both equations and taking the square root. This provides an expression for the intrinsic carrier density as a function of the effective density of states in the conduction and valence band, and the bandgap energy $E_g = E_c - E_v$.

$$n_i = \sqrt{N_c N_v}\, e^{-E_g/2kT} \quad \text{... (1.70)}$$

The temperature dependence of the intrinsic carrier density is dominated by the exponential dependence on the energy bandgap. In addition, one has to consider the temperature dependence of the effective densities of states and that of the energy bandgap. A plot of the intrinsic carrier density versus temperature is shown in Fig. 1.21.

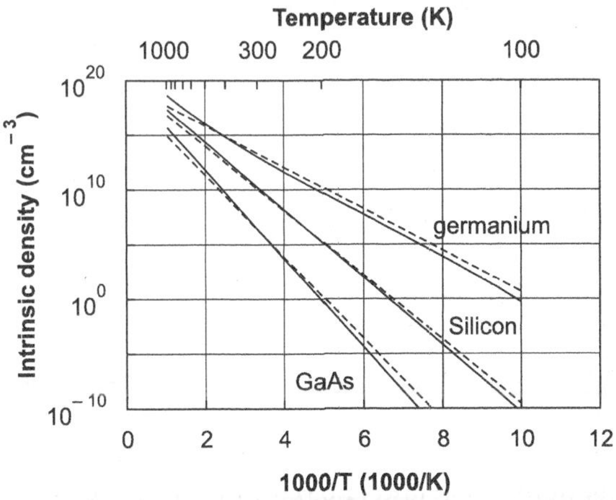

Fig. 1.21 *Intrinsic carrier density versus temperature in gallium arsenide (GaAs), silicon and germanium. Compared is the calculated density with (solid lines) and without (dotted lines) the temperature dependence of the energy bandgap*

Example. *Calculate the intrinsic carrier density in germanium, silicon and gallium arsenide at 300, 400, 500 and 600 K.*

Solution. The intrinsic carrier density in silicon at 300 K equals:

$$n_i\,(300\text{ K}) = \sqrt{N_c N_v}\,\exp\left(\frac{-E_g}{2kT}\right)$$

$$= \sqrt{2.81\times10^{19}\times1.83\times10^{19}}\,\exp\left(\frac{-1.12}{2\times0.0258}\right)$$

$$= 8.72\times10^{9}\text{ cm}^{-3}.$$

Similarly one finds the intrinsic carrier density for germanium and gallium arsenide at different temperatures, yielding:

	Germanium	*Silicon*	*Gallium Arsenide*
300 K	2.02×10^{13}	8.72×10^{9}	2.03×10^{6}
400 K	1.38×10^{15}	4.52×10^{12}	5.98×10^{9}
500 K	1.91×10^{16}	2.16×10^{14}	7.98×10^{11}
600 K	1.18×10^{17}	3.07×10^{15}	2.22×10^{13}.

1.9.3.2 Mass action law

Using the same approach as in section 1.9.3.1, one can prove that the product of the electron and hole density equals the square of the intrinsic carrier density for any non-degenerate semiconductor. By multiplying the expressions for the electron and hole densities in a non-degenerate semiconductor, as in equations (1.61) and (1.64), one obtains:

$$n_o \cdot p_o = N_c N_v\, e^{(E_v - E_c)/kT} = n_i^2 \qquad \text{... (1.71)}$$

This property is referred to as the mass action law. It is a powerful relation, which enables to quickly find the hole density if the electron density is known or vice versa. This relation is only valid for non-degenerate semiconductors in thermal equilibrium.

1.9.3.3 Intrinsic fermi energy

The above equations for the intrinsic electron and hole density can be solved for the intrinsic Fermi energy by taking the ratio of equations (1.69), yielding:

$$E_i = \frac{E_c + E_v}{2} + \frac{1}{2}kT\ln\left(\frac{N_v}{N_c}\right) \qquad \text{... (1.72)}$$

The intrinsic Fermi energy is typically close to the *midgap* energy, half way between the conduction and valence band edge. The intrinsic Fermi energy can also be expressed as a function of the effective masses of the electrons and holes in the semicondcutor. For this we use equations (1.62) and (1.65) for the effective density of states in the conduction and valence band, yielding:

$$E_i = \frac{E_c + E_v}{2} + \frac{3}{4} kT \ln\left(\frac{m_h^*}{m_e^*}\right) \qquad ...(1.73)$$

1.9.3.4 *Intrinsic material as reference*

Dividing the expressions for the carrier densities (1.61) and (1.64), by those for the intrinsic density (1.69) allows to write the carrier densities as a function of the intrinsic density and the intrinsic Fermi energy, E_i, or:

$$n_o = n_i\, e^{(E_F - E_i)/kT} \qquad ...(1.74)$$

and

$$p_o = n_i\, e^{(E_i - E_F)/kT} \qquad ...(1.75)$$

We will use primarily these two equations to find the electron and hole density in a semiconductor in thermal equilibrium. The same relations can also be rewritten to obtain the Fermi energy from either carrier density, namely:

$$E_F = E_i + kT \ln \frac{n_o}{n_i} \qquad ...(1.76)$$

and

$$E_F = E_i - kT \ln \frac{p_o}{n_i} \qquad ...(1.77)$$

1.9.4 Doped Semiconductors

Doped semiconductors are semiconductors, which contain impurities, foreign atoms incorporated into the crystal structure of the semiconductor. Either these impurities can be unintentional, due to lack of control during the growth of the semiconductor, or they can be added on purpose to provide free carriers in the semiconductor.

The generation of free carriers requires not only that impurities are present, but also that the impurities give off electrons to the conduction band in which case they are called *donors*. If they give off holes to the valence band, they are called *acceptors* (since they actually accept an electron from the filled valence band). The ionization of shallow donors and acceptors are illustrated by Fig. 1.22. Indicated are the donor and acceptor energies, E_d and E_a. The donor energy level is filled prior to ionization. Ionization causes the donor to be emptied, yielding an electron in the conduction band and a positively charged donor ion. The acceptor energy is empty prior to ionization. Ionization of the acceptor corresponds to the empty acceptor level being filled by an electron from the filled valence band. This is equivalent to a hole given off by the acceptor atom to the valence band.

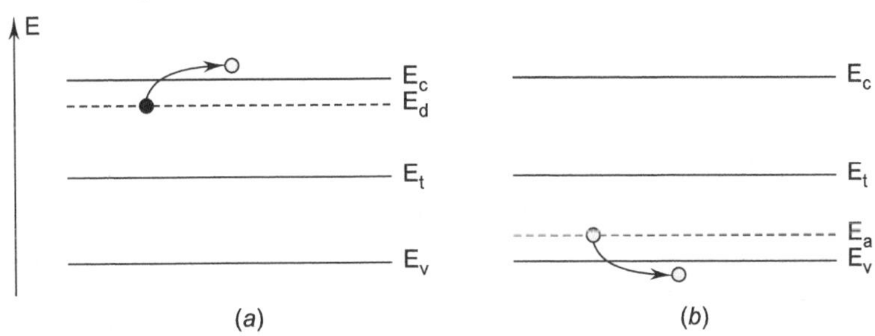

Fig. 1.22 *Ionization of (a) a shallow donor and (b) a shallow acceptor*

A semiconductor doped with impurities, which are ionized (meaning that the impurity atoms either have donated or accepted an electron), will therefore contain free carriers. *Shallow impurities* are impurities, which require little energy—typically around the thermal energy, kT, or less—to ionize. *Deep impurities* require energies much larger than the thermal energy to ionize so that only a fraction of the impurities present in the semiconductor contribute to free carriers. Deep impurities, which are more than five times the thermal energy away from either band edge, are very unlikely to ionize. Such impurities can be effective recombination centers, in which electrons and holes are captured and annihilate each other. Such deep impurities are also called *traps*.

Ionized donized donors provide free electrons in a semiconductor, which is then called *n-type*, while ionized acceptors provide free holes in a semiconductor, which we refer to as being a *p-type* semiconductor.

1.9.4.1 Dopants and impurities

The ionization of the impurities is dependent on the thermal energy and the position of the impurity level within the energy bandgap as described by the impurity distribution functions discussed in section 1.8.3.

Shallow impurities readily ionize so that the free carrier density equals the impurity concentration. For shallow donors this implies that the electron density equals the donor concentration, or:

$$n_o \cong N_d^+ = N_d \qquad \text{... (1.78)}$$

While for shallow acceptors the hole density equals the acceptor concentration, or:

$$p_0 \cong N_a^- = N_a \qquad \text{... (1.79)}$$

If a semiconductor contains both shallow donors and shallow acceptors it is called *compensated* since equal amounts of donor and acceptor atoms compensate each other, yielding no free carriers. The presence of shallow donors and shallow acceptors in a semiconductor cause the electrons given off by the donor atoms to fall into the acceptor state, which ionizes the acceptor atoms without yielding a free electron or hole. The resulting carrier density in compensated material is approximately equal to the difference between the donor and acceptor concentration if the donor concentration is larger, yielding n-type material, or:

$$n_o \cong N_d^+ - N_a^-, \text{ if } N_d^+ - N_a^- >> n_i \qquad \text{... (1.80)}$$

If the acceptor concentration is larger than the donor concentration, the hole density of the resulting p-type material equals the difference between the acceptor and donor concentration, or:

$$p_o \cong N_a^- - N_d^+, \text{ if } N_a^- - N_d^+ >> n_i \qquad \text{... (1.81)}$$

1.9.4.2 Ionization energy model

The energy required to remove an electron from a donor atom can be approximated using a hydrogen-like model. After all, the donor atom consists of a positively charged ion and an electron just like the proton and electron of the hydrogen atom. The difference however is that the average distance, r, between the electron and the donor ion is much larger since the electron occupies one of the outer orbitals. This is illustrated by Fig. 1.23.

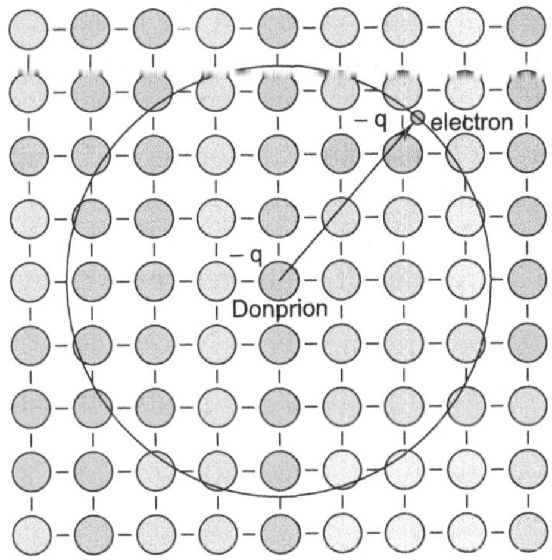

Fig. 1.23 *Trajectory of an electron bound to a donor ion within a semiconductor crystal. A 2-D square lattice is used for ease of illustration*

For shallow donors, this distance, *r*, is much larger than the inter-atomic spacing of the semiconductor crystal. The ionization energy, E_d, can be estimated by equation:

$$E_c - E_d = 13.6 \frac{m^*_{\text{cond}}}{m_0 \varepsilon_r^2} eV \quad \text{... (1.82)}$$

where m^*_{cond} is the effective mass for conductivity calculations and ε_r is the relative dielectric constant of the semiconductor. The ionization energy is calculated as the difference between the energy of a free electron and that of an electron occupying the lowest energy level, E_1.

Example. *Calculate the ionization energy for shallow donors and acceptors in germanium and silicon using the hydrogen-like model.*

Solution. Using the effective mass for conductivity calculations, one finds the ionization energy for shallow donors in germanium to be:

$$E_c - E_d = 13.6 \frac{m^*_{\text{cond}}}{m_0 \varepsilon_r^2} eV = 13.6 \frac{0.12}{(16)^2} eV = 6.4\, meV.$$

The calculated ionization energies for donors and acceptors in germanium and silicon are provided below.

	Germanium	Silicon
donors	6.4 meV	13.8 meV
acceptors	11.2 meV	20.5 meV

Note that the actual ionization energies differ from these values and depend on the actual donor atom.

1.9.4.3 Analysis of non-degenerately doped semiconductors

The calculation of the electron density stars by assuming that the semiconductor is neutral, so that there is no net charge in the material. This is a reasonable assumption since a net charge would result in an electric field. This electric field would move any mobile charge until it eliminates any charge imbalance.

The charge density in a semiconductor depends on the free electron and hole density and on the ionized impurity densities. Ionized donors, which have given off an electron, are positively charged. Ionized acceptors, which have accepted an electron, are negatively charged. The total charge density is therefore given by:

$$\rho = q\,(p_o - n_o + N_d^+ - N_a^-) = 0 \qquad \text{... (1.83)}$$

The hole concentration in thermal equilibrium can be written as a function of the electron density by using the mass action law (1.71). This yields the following relation between the electron density and the ionized impurity densities:

$$n_o = \frac{n_i^2}{n} + N_d^+ - N_a^- \qquad \text{... (1.84)}$$

Note that the use of the mass action law restricts the validity of this derivation to non-degenerate semiconductors as defined in section 1.70. Solving this quadratic equation yields a solution for the electron density, namely:

$$n_o = \frac{N_d^+ - N_a^-}{2} + \sqrt{\left(\frac{N_d^+ - N_a^-}{2}\right)^2 + n_i^2} \qquad \text{... (1.85)}$$

The same derivation can be repeated for holes, yielding:

$$p_o = \frac{N_a^- - N_d^+}{2} + \sqrt{\left(\frac{N_a^- - N_d^+}{2}\right)^2 + n_i^2} \qquad \text{... (1.86)}$$

The above expressions provide the free carrier densities for compensated semiconductors assuming that all donors and acceptors are ionized.

From the carrier densities, one then obtains the Fermi energies using equations (1.76) and (1.77) which are repeated below:

$$E_F = E_i + kT \ln \frac{n_o}{n_i} \qquad \text{... (1.87)}$$

or

$$E_F = E_i - kT \ln \frac{p_o}{n_i} \qquad \text{... (1.88)}$$

The calculated Fermi energies in n-type and p-type silicon as a function of doping density are shown in Fig. 1.24 for different temperatures:

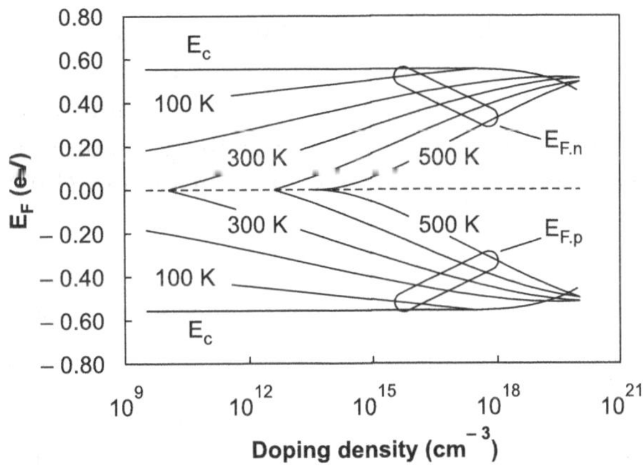

Fig. 1.24 *Fermi energy of n-type and p-type silicon, $E_{F,n}$ and $E_{F,p}$, as a function of doping density at 100, 200, 300, 400 and 500 K. Shown are the conduction and valence band edges, E_c and E_v. The midgap energy is set to zero*

Fig. 1.24 illustrates how the Fermi energies vary with doping density. The Fermi energy varies linearly, when plotting the density on a logarithmic scale, up to a doping density of 10^{18} cm^{-3}. This simple dependence requires that the semiconductor is neither intrinsic nor degenerate and that all the dopants are ionized. For compensated material, containing only shallow dopants, one uses the net doping density, $| N_d - N_a |$.

Example 1. *A germanium wafer is doped with a shallow donor density of $\frac{3n_i}{2}$. Calculate the electron and hole density.*

Solution. The electron density is obtained from equation (1.82) and the hole density is obtained using the mass action law:

Example 2. *A silicon wafer is doped with a shallow acceptor doping of 10^{16} cm^{-3}. Calculate the electron and hole density.*

Solution. Since the acceptor doping is much larger than the intrinsic density and much smaller than the effective density of states, the hole density equals: The electron density is then obtained using the mass action law.

1.9.4.4 General analysis

A more general analysis takes also into account the fact that the ionization of the impurities is not 100%, but instead is given by the impurity distribution functions provided in section 1.8.3.

The analysis again assumes that there is no net charge in the semiconductor (*charge neutrality*). This also means that the total density of positively charged particles (holes and ionized donors) must equals the total density of negatively charged particles (electrons and ionized acceptors) yielding:

$$p_o + N_d^+ = n_o + N_a^- \quad \text{... (1.89)}$$

The electron and hole densities are then written as a function of the Fermi energy. For a non-degenerate semiconductors one uses equations (1.60) and (1.63), while the ionized impurity densities equal the impurity density multiplied with the probability of occupancy for the acceptors and one minus the probability of occupancy for the donors. The Joyce-Dixon approximation, described in section (1.9.2.3) is used to calculate the degenerate carrier densities.

A graphical solution to equation (1.89) above can be obtained by plotting both sides of the equation as a function of the Fermi energy as illustrated in Fig. 1.25.

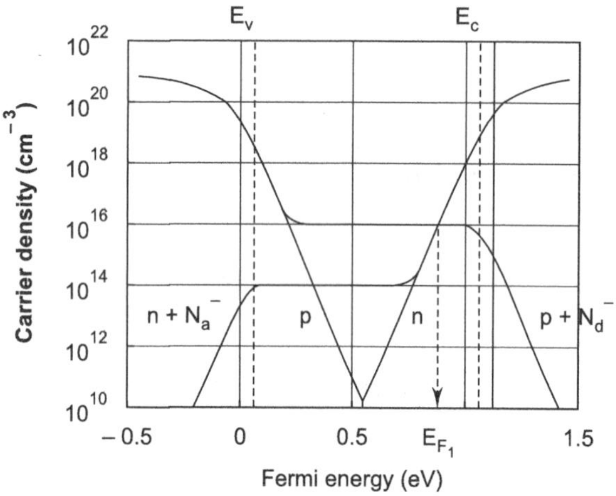

Fig. 1.25 *Graphical solution of the Fermi energy based on the general analysis. The value of the Fermi energy and the free carrier density is obtained at the intersection of the two curves, which represent the total positive and total negative charge in the semiconductor.* N_a *equals* 10^{16} cm^{-3} *and* N_d *equals* 10^{14} cm^{-3}

Fig. 1.25 shows the positive and negative charge densities as well as the electron and hole densities as a function of the Fermi energy. The dotted lines indicate the position of the acceptor and donor energies. The Fermi energy is obtained at the intersection of both curves as indicated by the arrow.

This graphical solution is a very useful tool to explore the Fermi energy as a function of the doping densities, ionization energies and temperature.

Operation of devices over a wide temperature range requires a detailed knowledge of the carrier density as a function of temperature. At intermediate temperatures the carrier density approximately equals the net doping, $| N_a - N_d |$. Semiconductors, which satisfy this condition, are also called extrinsic semiconductors. The free carrier density increases at high temperatures for which the intrinsic density approaches the net doping density and decreases at low temperatures due to incomplete ionization of the dopants. The carrier density and Fermi energy are shown in Fig. 1.26 for silicon doped with 10^{16} cm^{-3} donors and 10^{15} cm^{-3} acceptors:

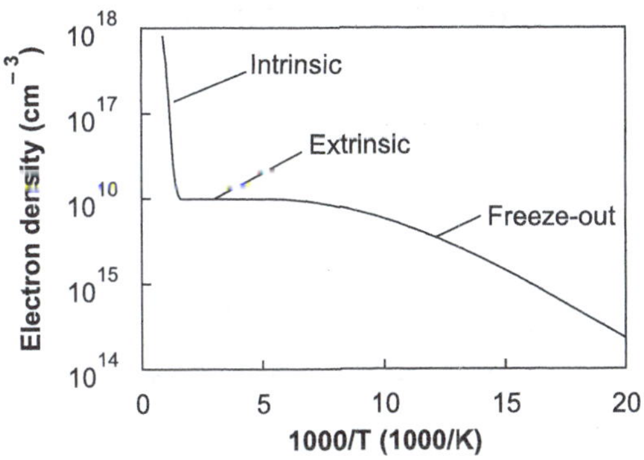

Fig. 1.26 *Electron density and Fermi energy as a function of temperature in silicon with* $N_d = 10^{16}$ *cm*$^{-3}$, $N_a = 10^{14}$ *cm*$^{-3}$ *and* $E_c - E_d = E_a - E_v = 50$ *meV. The activation energy at 70 K equals 27.4 meV*

At high temperatures, the carrier density equals the intrinsic carrier concentration, while at low temperatures the carrier density is dominated by the ionization of the donors.

The temperature dependence is related to an activation energy by fitting the carrier density versus $\frac{1}{T}$ on a semi-logarithmic scale to a straight line of the form n_o (T) = C exp $\left(\frac{-E_A}{kT}\right)$, where C is a constant. At high temperatures this activation energy equals half the bandgap energy of $E_A = \frac{E_g}{2}$.

The temperature dependence at low temperatures is somewhat more complex as it depends on whether or not the material is compensated. Fig. 1.26 was calculated for silicon containing both donors and acceptors. At 70 K the electron density is below the donor density but still larger than the acceptor density. Under such conditions the activation energy, E_A, equals half of the ionization energy of the donors of $\frac{(E_c - E_d)}{2}$. At lower temperatures where the electron density is lower than the acceptor density, the activation energy equals the ionization energy or $E_c - E_d$. This behavior is explained by the fact that the Fermi energy in compensated material is fixed at the donor energy. The donors levels are always partially empty as electrons are removed from the donor atoms to fill the acceptor energy levels. If the acceptor density is smaller than the electron density—as is true for uncompensated material—the Fermi energy does change with temperature and the activation energy approaches half of the ionization energy.

Lightly doped semiconductors suffer from freeze-out at relatively high temperature. Higher-doped semiconductors freeze-out at lower temperatures. Highly-doped semiconductors do not contain a single donor energy, but rather an impurity band which overlaps with the conduction or valence band. The overlap of the two bands results in free carriers even at zero Kelvin. Degenerately doping a semiconductor therefore eliminates freeze-out effects.

1.9.4.5 Temperature dependence of the carrier density in non-degenerate material

The analysis of non-degenerate semiconductors can be further expanded to include carrier freeze-out in the presence of donors as well as acceptors. Equation (1.89) can be written as:

$$p_o + \frac{N_d}{1+2\exp\frac{E_F - E_d}{kT}} = n_o + \frac{N_a}{1+4\exp\frac{E_a - E_F}{kT}} \qquad \text{... (1.90)}$$

For n-type material ($N_d > N_a$), this equation can be simplified to:

$$\frac{N_d}{1+\frac{n_o}{N^*}} = n_o + N_a \qquad \text{... (1.91)}$$

where the hole density, p_o, has been ignored and the acceptors are assumed to be fully ionized since the Fermi energy is expected to be above the energy of even the deepest acceptors. The density N^* is given by:

$$N^* = \frac{N_c}{2}\exp\frac{E_d - E_c}{kT} \qquad \text{... (1.92)}$$

The solution to the quadratic equation is:

$$n_o = \frac{N^* + N_a}{2} + \sqrt{\frac{(N^* + N_a)^2}{4} + N^*(N_d - N_a)} \qquad \text{... (1.93)}$$

Example: *4H-SiC is doped with 2×10^{17} cm^{-3} nitrogen donor atoms ($E_c - E_d = 90$ meV). Use $N_c = 4 \times 10^{20}$ cm^{-3}.*

(a) Calculate the electron density at 300 K.

(b) Calculate the hole density at 300 K after adding 2×10^{18} cm^{-3} aluminium acceptor atoms ($E_a - E_c = 220$ meV). Use $N_v = 1.6 \times 10^{20}$ cm^{-3}.

Solution.

(*a*) First we calculate N^*

$$N^* = \frac{N_c}{2}\exp\frac{E_d - E_c}{kT} = 6.16 \times 10^{17}\ \text{cm}^{-3}$$

The free electron density is then obtained from:

$$n_o = -\frac{N^*}{2} + \sqrt{\frac{(N^*)^2}{4} + N^* N_d} = 4.3 \times 10^{16}\ \text{cm}^{-3}$$

As a result 21.5% of the donors are ionized

(*b*) Since we are now dealing with p-type material we have to recalculate N^*.

$$N^* = \frac{N_v}{4}\exp\frac{E_v - E_c}{kT} = 8.08 \times 10^{16}\ \text{cm}^{-3}$$

where the factor 4 is due to the doubly degenerate valence band. The free hole density is then obtained from:

$$p_o = -\frac{N^* + N_d}{2} + \sqrt{\frac{(N^* + N_d)^2}{4} + N^*(N_a - N_d)}$$
$$= 5.52 \times 10^{16}\ \text{cm}^{-3}$$

which is identical to (1.92) except that the donor and acceptor densities have been exchanged. Only 2.76% of the acceptors are ionized while all donors are ionized.

1.9.5 Non-equilibrium Carrier Densities

Up until now, we have only considered the thermal equilibrium carrier densities, n_o and p_o. However, most devices of interest are not in thermal equilibrium. Keep in mind that a constant ambient constant temperature is not a sufficient condition for thermal equilibrium. In fact, applying a non-zero voltage to a device or illuminating it with light will cause a non-equilibrium condition, even if the temperature is constant.

To describe a system that is not in thermal equilibrium we assume that each of the carrier distributions is still in equilibrium with itself. Such assumption is justified on the basis that electrons readily interact with each other and interact with holes only on a much longer time scale. As a result the electron density can still be calculated using the Fermi-Dirac distribution function, but with a different value for the Fermi energy. The total carrier density for a non-degenerate semiconductor is then described by:

$$n = n_o + \delta n + n_i \exp\left(\frac{F_n - E_i}{kT}\right) \qquad \text{... (1.94)}$$

Where ∂n is the *excess electron density* and F_n is the *quasi-Fermi energy* for the electrons. Similarly, the hole density can be expressed as:

$$p = p_o + \delta p = n_i \exp\left(\frac{E_i - E_F}{kT}\right) \qquad \text{... (1.95)}$$

Where ∂p is the excess hole density and F_p is the *quasi-Fermi energy* for the holes.

Example. *A piece of germanium doped with 10^{16} cm^{-3} shallow donors is illuminated with light generating 10^{15} cm^{-3} excess electrons and holes. Calculate the quasi-Fermi energies relative to the intrinsic energy and compare it to the Fermi energy in the absence of illumination.*

Solution. The carrier densities when illuminating the semiconductor are:

$$n = n_o + \delta n = 10^{16} + 10^{15} = 1.1 \times 10^{16}\ \text{cm}^{-3}$$
$$p = p_o + \delta p \cong 10^{15}\ \text{cm}^{-3}$$

and the quasi-Fermi energies are:

$$F_n - E_i = kT \ln \frac{n}{n_i} = 0.0259 \times \ln \frac{1.1 \times 10^{16}}{2 \times 10^{13}} = 163\ \text{meV}.$$

$$F_p - E_i = -kT \ln \frac{p}{n_i} = 0.0259 \times \ln \frac{1 \times 10^{15}}{2 \times 10^{13}} = -101\ \text{meV}$$

For comparison, the Fermi energy in the absence of light equals

$$E_F - E_i = kT \ln \frac{n_o}{n_i} = 0.0259 \times \ln \frac{10^{16}}{2 \times 10^{13}} = 161 \text{ meV}$$

which is very close to the quasi-Fermi energy of the majority carriers.

QUESTIONS

- How does the conductivity of a solid depend on whether the energy bands are completely filled, partially filled or empty ? How does the existence of overlapping bandgaps affect the conductivity ?
- Why does a completely filled band not contribute the conductivity of a solid ?
- Explain physically why the bandgap of a semiconductor decreases with temperature.
- What are holes ? Carefully justify your definition.
- What is a state ?
- How many states are there in 1 micron sized cube for which an electron has a kinetic energy less than 1 eV ? Treat the electron as a free electron confined to a box with infinite potential walls.
- What is the physical meaning of the Fermi energy ?
- What is the value of the Fermi function at an energy, which is 3 kT larger/lower than the Fermi energy ?
- What is the basic assumption used in statistical thermodynamics when calculating the probability distribution functions ?
- What are the two boundary conditions used to find the possible ways to fill energy levels with electrons.
- What is an intrinsic semiconductors ? What is the hole density in an intrinsic semiconductor?
- Why is the product of the electron and hole density in a non-degenerate semiconductor constant rather than for instance the sum ? This relationship is also referred to as the mass-action law. Why ?
- Define a non-degenerate semiconductor. Why do we need this concept ?
- What is the difference between a doped semiconductor and an extrinsic semiconductor ?
- Describe the temperature dependence of the carrier density in a semiconductor. Identify the three regions and explain what happens by indicating the filled and empty states on an energy band diagram. Do this for n-type, p-type and compensated material.

2

Carrier Transport

2.1 INTRODUCTION

Any motion of free carriers in a semiconductor leads to a current. This motion can be caused by an electric field due to an externally applied voltage, since the carriers are charged particles. We will refer to this transport mechanism as carrier drift. In addition, carriers also move from regions where the carrier density is high to regions where the carrier density is low. This carrier transport mechanism is due to the thermal energy and the associated random motion of the carriers. We will refer to this transport mechanism as carrier diffusion. The total current in a semiconductor equals the sum of the drift and the diffusion current.

As one applies an electric field to a semiconductor, the electrostatic force causes the carriers to first accelerate and then reach a constant average velocity, v, due to collisions with impurities and lattice vibrations. The ratio of the velocity to the applied field is called the mobility. The velocity saturates at high electric fields reaching the saturation velocity. Additional scattering occurs when carriers flow at the surface of a semiconductor, resulting in a lower mobility due to surface or interface scattering mechanisms.

Diffusion of carriers is obtained by creating a carrier density gradient. Such gradient can be obtained by varying the doping density in a semiconductor or by applying a thermal gradient.

Both carrier transport mechanisms are related since the same particles and scattering mechanisms are involved. This leads to a relationship between the mobility and the diffusion constant called the Einstein relation.

2.2 CARRIER DRIFT

The motion of a carrier drifting in a semiconductor due to an applied electric field, ε, is illustrated in Fig. 2.1. The field causes the carrier to move with a velocity, v.

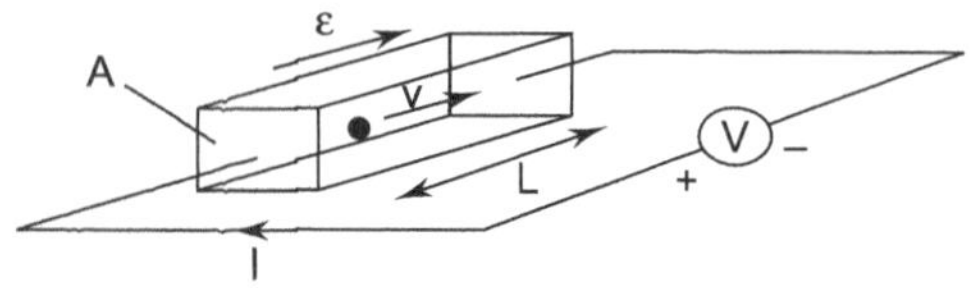

Fig. 2.1 *Drift of a carrier due to an applied electric field*

Assuming that all the carriers in the semiconductor move with the same velocity, the current can be expressed as the total charge in the semiconductor divided by the time needed to travel from one electrode to the other, or :

$$I = \frac{Q}{t_r} = \frac{Q}{L/v} \quad \text{... (2.1)}$$

where t_r is the transit time of a particle, traveling with velocity v, over the distance L. The current density, J, can then be rewritten as a function of the charge density, J :

$$\vec{J} = \frac{Q}{AL}\vec{v} = \rho\vec{v} = qn\vec{v} \quad \text{... (2.2)}$$

If the carriers are negatively charged electrons, the current density equals :

$$\vec{J} = -qn\vec{v} \quad \text{... (2.2 (a))}$$

while for positively charged holes it is :

$$\vec{J} = qp\vec{v} \quad \text{... (2.2 (b))}$$

Where n and p are the electron and hole density in the semiconductor.

It should be noted that carriers do not follow a straight path along the electric field lines. Instead they bounce around in the semiconductor and constantly change direction and velocity due to scattering. This behavior occurs even when no electric field is applied and is due to the thermal energy of the carriers. Thermodynamics teaches us that electrons in a non-degenerate and non-relativistic electron gas have a thermal energy of $\frac{kT}{2}$ per particle per degree of freedom. A typical thermal velocity at room temperature is around 10^7 cm/s, which exceeds the typical drift velocity in semiconductors. The carrier motion in the semiconductor in the absence and in the presence of an electric field can therefore be visualized as in Fig. 2.2.

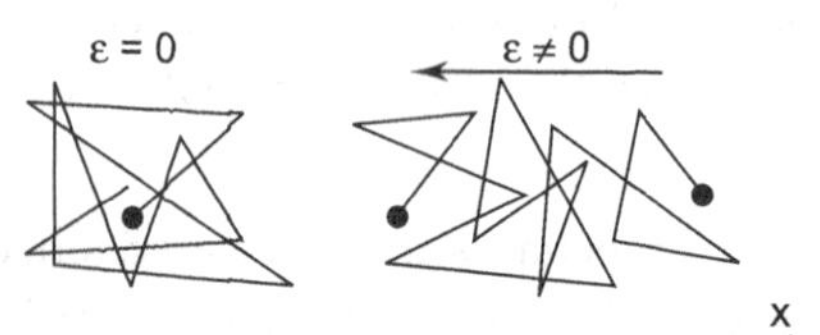

Fig. 2.2 *Random motion of carriers in a semiconductor with and without an applied electric field*

In the absence of an applied electric field, the carrier exhibits random motion and the carriers love quickly through the semiconductor and frequently change direction. When an electric field is applied, the random motion still occurs but in addition, there is on average a net motion along the direction of the field. Due to their different electronic charge, holes move on average in the direction of the applied field, while electrons move in the opposite direction.

We now analyze the carrier motion considering only the average velocity, $\langle \vec{v} \rangle$ of the carriers. Applying Newton's law, we state that the acceleration of the carriers is proportional to the applied force :

$$\vec{F} = m\vec{a} = m\frac{d\langle \vec{v} \rangle}{dt} \qquad \text{... (2.3)}$$

The force consists of the difference between the electrostatic force and the scattering force due to the loss of momentum at the time of scattering. This scattering force equals the momentum divided by the average time, τ_c, between scattering events or collisions, so that :

$$\vec{F} = q\vec{\varepsilon} - \frac{m\langle \vec{v} \rangle}{\tau_c} \qquad \text{... (2.4)}$$

where a particle with charge q was assumed.

Combining both relations yields an expression for the average particle velocity :

$$q\vec{\varepsilon} = m\frac{d\langle \vec{v} \rangle}{dt} + \frac{m\langle \vec{v} \rangle}{\tau_c} \qquad \text{... (2.5)}$$

We now consider only the steady state situation in which the particle has already accelerated and has reached a constant average velocity. Under such conditions, the velocity is proportional to the applied electric field and we define the mobility as the velocity to field ratio :

$$\mu = \frac{\Delta |\vec{v}|}{|\vec{\varepsilon}|} = \frac{q\tau_c}{m} \qquad \text{... (2.6)}$$

The mobility of a particle in a semiconductor is therefore expected to be large if its mass is small and the time between scattering events is large.

The drift current, described by (2.2), can then be rewritten as a function of the mobility, yielding:

$$\vec{J}_n = qn\mu_n\vec{\varepsilon} \qquad \text{... (2.7)}$$

For electrons and

$$\vec{J}_p = qp\mu_p\vec{\varepsilon} \qquad \text{... (2.8)}$$

For holes

Throughout this derivation, we simply considered the mass, m, of the particle. However, in order to incorporate the effect of the periodic potential of the atoms in the semiconductor we must use the effective mass, m^*, rather than the free particle mass :

$$\mu = \frac{q\tau_c}{m^*} \qquad \text{... (2.9)}$$

Example. *Electrons in undoped gallium arsenide have a mobility of 8,800 cm²/V-s. Calculate the average time between collisions. Calculate the distance traveled between two collisions (also called the mean free path). Use an average velocity of 10^7 cm/s.*

Solution. The collision time, τ_c, is obtained from :

$$\tau_c = \frac{\mu_n m_e^*}{q} = \frac{0.88 \times 0.067 \times 9.1 \times 10^{-31}}{1.6 \times 10^{-19}} = 0.34 \text{ ps}$$

where the mobility was first converted in MKS units.

The mean free path, l, equals :

$$l = v_{\text{average}}\, \tau_c = 10^7 \times 0.34 \times 10^{-12} = 34 \text{ nm}.$$

2.2.1 Scattering Phenomena

2.2.1.1 Impurity scattering

Impurities are foreign atoms in the solid, which are efficient scattering centres especially when they have a net charge. Ionized donors and acceptors in a semiconductor are a common example of such impurities. The amount of scattering due to electrostatic forces between the carrier and the ionized impurity depends on the interaction time and the number of impurities. Larger impurity concentrations result in a lower mobility. The dependence on the interaction time helps to explain the temperature dependence. The interaction time is directly linked to the relative velocity of the carrier and the impurity, which is related to the thermal velocity of the carriers. The thermal velocity increases with the ambient temperature so that the interaction time decreases. Thereby, the amount of scattering decreases, resulting in a mobility increase with temperature. To first order, the mobility due to impurity scattering is proportional to $\frac{T^{3/2}}{N_1}$, where N_1 is the density of charged impurities.

2.2.1.2 Lattice scattering

Scattering by lattice waves includes the absorption or emission of either acoustical or optical phonons. These phonons represent quanta of mechanical waves that travel through the semiconductor crystal. Since the density of phonons in a solid, increases with temperature, the scattering time due to this mechanism will decrease with temperature as will the mobility. Theoretical calculations reveal that the mobility in non-polar semiconductors, such as silicon and germanium, is dominated by *acoustic* phonon interaction. The resulting mobility is expected to be proportional to $T^{-3/2}$, while the mobility due to *optical* phonon scattering only is expected to be proportional to $T^{-1/2}$. Experimental values of the temperature dependence of the mobility in germanium, silicon and gallium arsenide are provided in Table 2.1.

Table 2.1 Temperature dependence of the mobility in germanium, silicon and gallium arsenide due to phonon scattering

	Germanium	*Silicon*	*Gallium Arsenide*
Electron mobility	$\propto T^{-1.7}$	$\propto T^{-2.4}$	$\propto T^{-1.0}$
Hole mobility	$\propto T^{-2.3}$	$\propto T^{-2.2}$	$\propto T^{-2.1}$

2.2.1.3 Surface scattering

The surface and interface mobility of carriers is affected by the nature of the adjacent layer or surface. Even if the carrier does not transfer into the adjacent region, its wavefunction does extend over 1 to 10 nanometer, so that there is a non-zero probability that the particle is in the adjacent region. The net mobility is then a combination of the mobility in both layers. For carriers in the inversion layer of a MOSFET, one finds that the mobility can be up to three times lower than the bulk value. This is due to the distinctly lower mobility of electrons in the amorphous silicon oxide. The presence of charged surface states further reduces the mobility just as ionized impurities would.

2.3 CARRIER MOBILITY

2.3.1 Doping Dependence

The mobility of electrons and holes in silicon at room temperature is shown in Fig. 2.3.

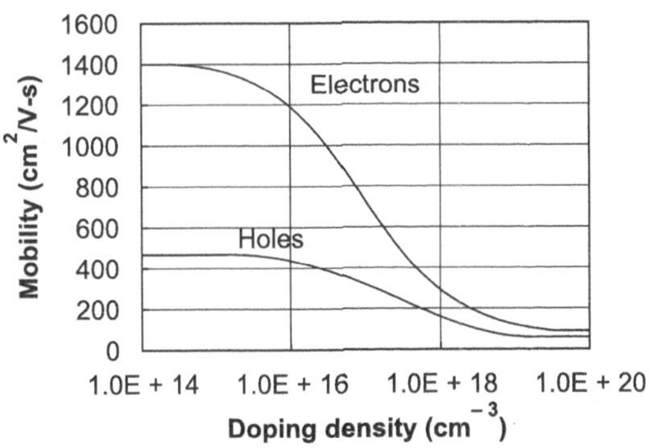

Fig. 2.3 *Electron and hole mobility versus doping density for silicon*

The electron mobility and hole mobility have a similar doping dependence : For low doping concentrations, the mobility is almost constant and is primarily limited by phonon scattering. At higher doping concentrations, the mobility decreases due to ionized impurity scattering with the ionized doping atoms. The actual mobility also depends on the type of dopant. Fig. 2.3 is for phosphorous and boron doped silicon.

Note that the mobility is linked to the total number of ionized impurities or the *sum* of the donor and acceptor densities. The free carrier density, as described in section 1.9.4.1 is to first order related to the *difference* between the donor and acceptor concentration.

The minority carrier mobility also depends on the total impurity density. The minority-carrier mobility can be approximated by the majority-carrier mobility in a material with the same number of impurities.

The mobility at a particular doping density is obtained from the following empiric expression :

$$\mu = \mu_{min} + \frac{\mu_{max} - \mu_{min}}{1 + \left(\frac{N}{N_r}\right)^{\alpha}} \qquad \text{... (2.10)}$$

where μ_{min}, μ_{max}, α and N_r are fit parameters. These parameters for arsenic, phosphorous and boron doped silicon are provided in Table 2.2.

Table 2.2 Parameters for calculation of the mobility as a function of the doping density

	Arsenic	*Phosphorous*	*Boron*
μ_{min} (cm^2/V-s)	52.2	68.5	44.9
μ_{max} (cm^2/V-s)	1417	1414	470.5
N_r (cm^{-3})	9.68×10^{16}	9.20×10^{16}	2.23×10^{17}
α	0.68	0.711	0.719

The resulting mobilities in units of cm^2/V-s are listed for different doping densities in Table 2.3.

Table 2.3 Mobility in silicon for different dopants and doping densities

N	*Arsenic*	*Phosphorous*	*Boron*
10^{15} cm^{-3}	1359	1362	462
10^{16} cm^{-3}	1177	1184	429
10^{17} cm^{-3}	727	721	317
10^{18} cm^{-3}	284	277	153
10^{19} cm^{-3}	108	115	71

2.4 CONDUCTIVITY AND RESISTIVITY

The conductivity of a material is defined as the current density divided by the applied electric field. Since the current density equals the product of the charge of the mobile carriers, their density and velocity as described by equations (2.2(*a*)) and (2.2(*b*)), it can be expressed as a function of the electric field using the mobility. To include the contribution of electrons as well as holes to the conductivity, we add the current density due to holes to that of the electrons, or :

$$J = qnv_e + qpv_h = q\,(n \,.\, \mu_n + p \,.\, \mu_p)\,\varepsilon \qquad \text{... (2.11)}$$

The conductivity due to electrons and holes is then obtained from :

$$\sigma = \frac{\Delta J}{\varepsilon} = q\left(n\mu_n + p\mu_p\right) \qquad \text{... (2.12)}$$

The resistivity is defined as the inverse of the conductivity, namely :

$$\rho = \frac{1}{\sigma} = \frac{1}{q\left(\mu_n n + \mu_p p\right)} \qquad \text{... (2.13)}$$

The resulting resistivity as calculated with the expression above is shown in Fig. 2.4.

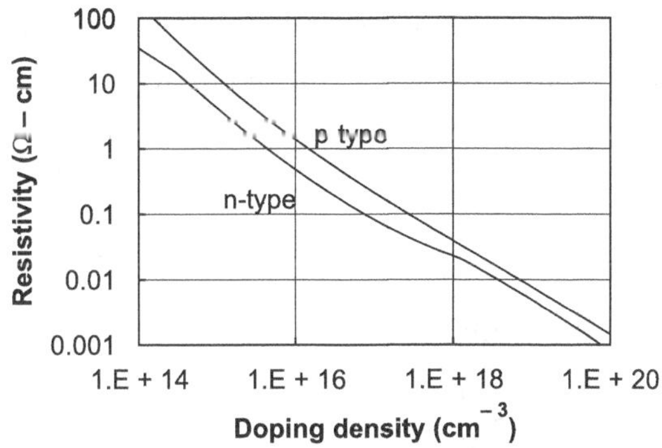

Fig. 2.4 *Resistivity of n-type and p-type silicon versus doping density.*

2.5 SHEET RESISTANCE

The sheet resistance concept is used to characterize both wafers and thin doped layers, since it is typically easier to measure the sheet resistance rather than the resistivity of the material. The sheet resistance of a uniformly-doped layer with resistivity, ρ, and thickness, *t*, is given by their ratio :

$$R_s = \frac{\rho}{t} \qquad \text{... (2.14)}$$

While the unit of the sheet resistance is Ohms, one refers to it as Ohms per square. This nomenclature comes in handy when the resistance of a rectangular piece of material with length, L, and width W must be obtained. It equals the product of the sheet resistance and the number of squares or :

$$R = R_s \frac{L}{W} \qquad \text{... (2.15)}$$

where the number of squares equals the length divided by the width. Fig. 2.5 provides, as an example, the sheet resistance of a 14 mil thick silicon wafer for both *n*-type and *p*-type silicon.

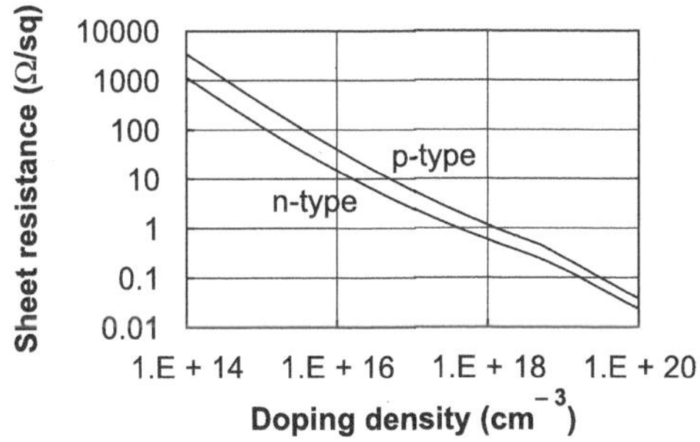

Fig. 2.5 *Sheet resistance of a 14 mil thick n-type and p-type silicon wafer versus doping density*

Example. *A piece of silicon doped with arsenic ($N_d = 10^{17}$ cm^{-3}) is 100 μm long, 10 μm wide and 1 μm thick. Calculate the resistance of this sample when contacted one each end.*

Solution. The resistivity of the silicon equals :

$$\rho = \frac{1}{qn\mu_n} = \frac{1}{1.6 \times 10^{-19} \times 10^{17} \times 727} = 0.086\ \Omega\ \text{cm}$$

where the mobility was obtained from Table 2.3.

The resistance then equals :

$$R = \rho\frac{L}{Wt} = 0.086 \times \frac{100 \times 10^{-4}}{10 \times 10^{-4} \times 10^{-4}} = 8.6\ k\Omega$$

An alternate approach is to first calculate the sheet resistance, R_s :

$$R_s = \frac{\rho}{t} = \frac{0.086}{10^{-4}} = 860\ \Omega/\text{square}$$

From which one then obtains the resistance :

$$R = R_s\frac{L}{W} = 860 \times \frac{100 \times 10^{-4}}{10 \times 10^{-4}} = 8.6\ k\Omega.$$

2.6 VELOCITY SATURATION

The linear relationship between the average carrier velocity and the applied field breaks down when high fields are applied. As the electric field is increased, the average carrier velocity and the average carrier energy increases as well. When the carrier energy increases beyond the optical phonon energy, the probability of emitting an optical phonon increases abruptly. This mechanism causes the carrier velocity to saturate with increasing electric field. For carriers in silicon and other materials, which do not contain accessible higher bands, the velocity versus field relation increases monotonically as shown in Fig. 2.6a.

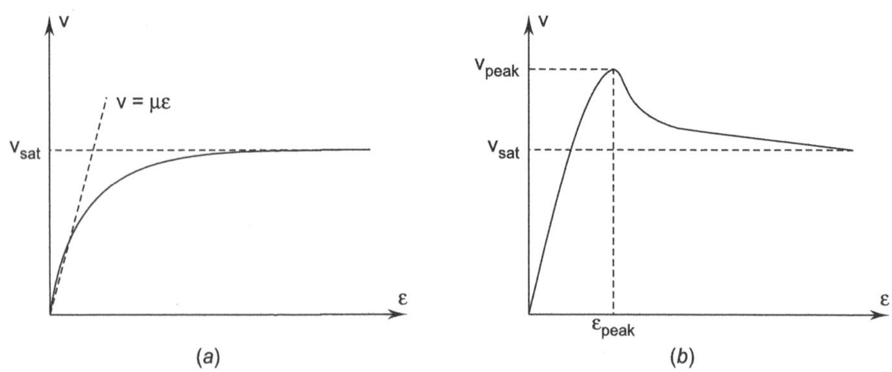

Fig. 2.6 *Velocity-field relation for (a) materials without accessible higher bands and (b) materials with an accessible higher band*

For those materials the velocity-field relation can be described by :

$$v(\varepsilon) = \frac{\mu\varepsilon}{1 + \dfrac{\mu\varepsilon}{v_{sat}}} \quad \text{... (2.16)}$$

The maximum obtainable velocity, v_{sat}, is referred to as the saturation velocity.

A first estimate of the saturation velocity, v_{sat}, of a material can be obtained by calculating the carrier velocity when it reaches an energy equal to the optical phonon energy.

$$v_{sat} = \sqrt{\frac{2E_{phonon}}{m^*}} \quad \text{... (2.17)}$$

While this equation can at best provide a rough estimate of the actual saturation velocity, it does indicate the right trend. Materials with a small effective mass and high optical phonon energy are more likely to have a high saturation velocity. Materials with multiple band minima can have a rather low saturation velocity, relative to the peak velocity, if the carriers in the higher minima have a larger effective mass.

2.7 CARRIER DIFFUSION

Carrier diffusion is due to the thermal energy, kT, which causes the carriers to move at random even when no field is applied. This random motion does not yield a net flow of carriers nor does it yield a net current in material with a uniform carrier density as any carrier which leaves a specific location is on average replace by another one. However, if a carrier gradient is present, the diffusion process will even out the carrier density variations: Carriers *diffuse* from regions where the density is high to regions where the density is low. The diffusion process is not unlike the motion of sand on a vibrating table; hills as well as valleys are smoothed out over time.

In this section we will first derive the expression for the current due to diffusion and then combine it with the drift current to obtain the total drift-diffusion current.

2.7.1 A Diffusion Current

The derivation is based on the basic notion that carries at non-zero temperature (Kelvin) have an additional thermal energy, which equals $\frac{kT}{2}$ per degree of freedom. It is the thermal energy, which drives the diffusion process. At T = 0 K there is no diffusion.

The reader should recognize that the random nature of the thermal energy would normally require a statistical treatment of the carrirs. Instead we will use average values to describe the process. Such approach is justified on the basis that a more elaborate statistical approach yields the same results. To further simplify the derivation, we will derive the diffusion current for a one-dimensional semiconductor in which carriers can only move along one direction.

We now introduce the average values of the variables of interest, namely the thermal velocity, v_{th}, the collision time, τ_c, and the mean free path, l. The thermal velocity is the average velocity of the carriers going in the positive or negative direction. The collision time is the time during which carriers will move with the same velocity before a collision occurs with an atom or with another carrier. The mean free path is the average length a carrier will travel between collisions. These three averages are related by :

$$v_{th} = \frac{l}{r} \qquad \text{... (2.18)}$$

Consider now the situation illustrated with Figure 2.7.

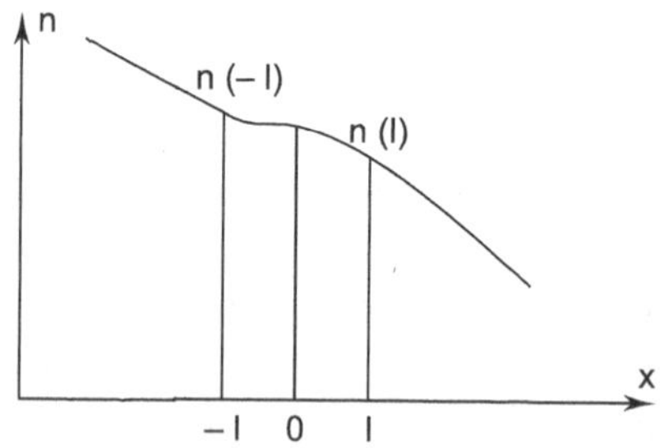

Fig. 2.7 *Carrier density profile used to derive the diffusion current expression*

Shown is a variable carrier density, $n(x)$. Of interest are the carrier densities which are one mean free path away from $x = 0$, since the carriers, which will arrive at $x = 0$ originate either at $x = -l$ or $x = l$. The flux at $x = 0$ due to carriers that originate at $x = -1$ and move from left to right equals :

$$\Phi_{n,\text{left}\rightarrow\text{right}} = \frac{1}{2} v_{th}\, n(x = -l) \qquad \text{... (2.19)}$$

where the factor $\frac{1}{2}$ is due to the fact that only half of the carriers move to the left while the other half moves to the right. The flux at $x = 0$ due to carriers that originate at $x = +l$ and move from right to left, equals :

$$\Phi_{n,\text{right}\rightarrow\text{left}} = \frac{1}{2} v_{th}\, n(x = l) \qquad \text{... (2.20)}$$

The total flux of carriers moving from left to right at $x = 0$ therefore equals :

$$\Phi_n = \Phi_{n,\text{left}\rightarrow\text{right}} - \Phi_{n,\text{right}\rightarrow\text{left}} = \frac{1}{2} v_{th} [n(x = -1) - n(x = l)] \qquad \text{... (2.21)}$$

Where the flux due to carriers moving from right to left is subtracted from the flux due to carriers moving from left to right. Given that the mean free path is small we can write the difference in densities divided by the distance between $x = -l$ and $x = l$ as the derivative of the carrier density :

$$\Phi_n = -lv_{th} \frac{n(x = l) - n(x = -l)}{2l} = -lv_{th} \frac{dn}{dx} \qquad \text{... (2.22)}$$

The electron diffusion current equals this flux times the charge of an electron, or :

$$J_n = -q\Phi_n = ql v_{th} \frac{dn}{dx} \quad \text{... (2.23)}$$

We now replace the product of the thermal velocity and the mean free path by a single parameter, namely the diffusion constant, D_n.

$$J_n = qD_n \frac{dn}{dx} \quad \text{... (2.24)}$$

Repeating the same derivation of holes yields :

$$J_p = -qD_p \frac{dp}{dx} \quad \text{... (2.25)}$$

We now further explore the relation between the diffusion constant and the mobility. At first, it seems that there should be no relation between the two since the driving force is distinctly different: Diffusion is caused by thermal energy while an externally applied field causes drift. However, one essential parameter in the anlaysis, namely the collision time, τ_c, should be independent of what causes the carrier motion.

We now combine the relation between the velocity, mean free path and collision time,

$$v_{th} = \frac{l}{r} \quad \text{... (2.26)}$$

with the result from thermodynamics, stating that electrons carry a thermal energy which equals $\frac{kT}{2}$ for each degree of freedom. Applied to a one-dimensional situation, this leads to :

$$\frac{kT}{2} = \frac{m^* v_{th}^2}{2} \quad \text{... (2.27)}$$

We now use these relations to rewrite the product of the thermal velocity and the mean free path as a function of the carrier mobility :

$$l v_{th} = \frac{m^* v_{th}^2}{q} \frac{q r}{m^*} = \frac{kT}{q} \mu \quad \text{... (2.28)}$$

Using the definition of the diffusion constant we then obtain the following expressions which are often referred to as the Einstein relations :

$$D_n = \mu_n \frac{kT}{q} = \mu_n V_t \quad \text{... (2.29)}$$

$$D_p = \mu_p \frac{kT}{q} = \mu_p V_t \quad \text{... (2.30)}$$

Example. *The hole density in an n-type silicon wafer ($N_d = 10^{17}$ cm^{-3}) decreases linearly from 10^{14} cm^{-3} to 10^{13} cm^{-3} between x = 0 and x = 1 mm. Calculate the hole diffusion current density.*

Solution. The hole diffusion current density equals :

$$J_p = qD_p \frac{dp}{dx} = 1.6 \times 10^{-19} \times 8.2 \times \frac{9 \times 10^{13}}{10^{-4}} = 1.18 \text{ A/cm}^2$$

where the diffusion constant was calculated using the Einstein relation :

$$D_p = V_t \mu_p = 0.0259 \times 317 = 8.2 \text{ cm}^2/\text{s}$$

and the hole mobility in the *n*-type wafer was obtained from Table 2.3 as the hole mobility in a *p*-type material with the same doping density.

2.7.2 Einstein Relation

The semiconductor devices in general operate under non-equilibrium conditions.

For example, when a bar of *n*-type germanium acting as a photodetecting device is illuminated with light of sufficient energy, excess charge carriers are produced in the exposed region of the material. Hence, the semiconductor is no longer in thermal equilibrium. This leads to a diffusion currents creating an internal electric field and a drift current balancing the diffusion component.

i.e., $$(\Delta n)\, e\, E\, \mu_e = e\, D_n \frac{\delta (4n)}{dx}$$

K = Boltman Constant

T = Absolute temp.

The force F on excess carriers restoring equilibrium is given by the product of excess charge and electric field.

$$F = (\Delta n)\, eE = (d\, D_n / \mu_e) \frac{\delta (4n)}{dx}$$

According to the kinetic theory of gases, if the molecular concentration of a gas increases by Δn, the gas pressure increases by $\Delta n\, kT$.

The force corresponding to the pressure gradient is given by $kT \frac{\delta}{\delta x} (\Delta n) = F$.

By making an analogy between the excess carriers in a semiconductors and gas molecules in a gas container.

$$kT = e\, D_n / \mu_e$$

$$\therefore \quad D_n = \frac{\mu_e\, kT}{e}$$

This relation between diffusion coefficient and mobility of a charge carrier is termed Einstein relation.

This relation is given for holes by

$$D_p = \mu_n kT/e$$

Hence, $$\frac{D_n}{D_p} = \frac{\mu_e}{\mu_n}.$$

* Value of D :-

Si : $D_n = 3.4 \times 10^{-3} \text{ m}^2/\text{s}$

$$D_p = 1.3 \times 10^{-3}\ m^2/s$$

Ge:

$$D_n = 3.4 \times 10^{-3}\ m^2/s$$

$$D_p = 1.3 \times 10^{-3}\ m^2/s.$$

2.7.3 Total Current

The total electron current is obtained by adding the current due to diffusion to the drift current, resulting in :

$$J_n = qn\mu_n\,\varepsilon + qD_n \frac{dn}{dx} \qquad \text{... (2.31)}$$

and similarly for holes :

$$J_p = qp\mu_p\,\varepsilon - qD_r \frac{dp}{dx} \qquad \text{... (2.32)}$$

The total current is the sum of the electron and hole current densities multiplied with the area, A, perpendicular to the direction of the carrier flow :

$$I_{total} = A\,(J_n + J_p) \qquad \text{... (2.33)}$$

2.8 THE HALL EFFECT

The Hall effect describes the behavior of the free carriers in a semiconductor when applying an electric as well as a magnetic field. The experimental set up shown in Fig. 2.8, depicts a semiconductor bar with a rectangular cross section and length L. A voltage V_x is applied between the two contacts, resulting in a field along the x-direction. The magnetic field is applied in the z-direction.

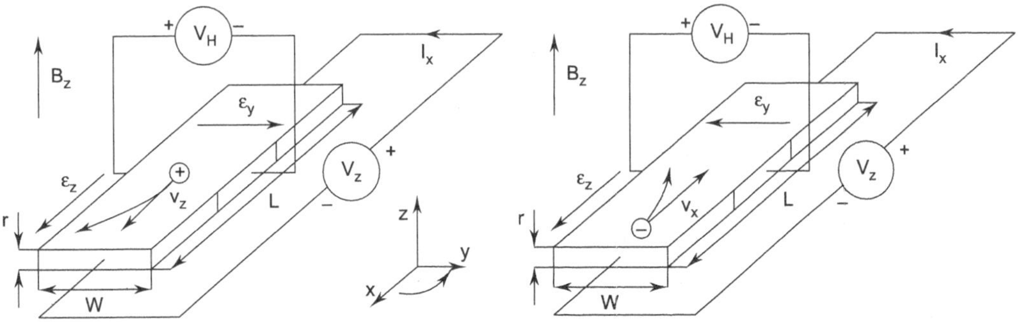

Fig. 2.8 *Hall set up and carrier motion for (a) holes and (b) electrons.*

As shown in Fig. 2.8*a*, the holes move in the positive x-direction. The magnetic field causes a force to act on the mobile particles in a direction dictated by the right hand rule. As a result there is a force, F_y, along the positive y-direction, which moves the holes to the right. In steady state this force is balanced by an electric field, ε_y, so that there is no net force on the holes. As a result there is a voltage across the sample, which can be measured with a high-impedance voltmeter. This voltage, V_H, is called the Hall voltage. For the sign convention shown in 2.8, the Hall voltage is positive for holes.

The behavior of electrons is shown in 2.8b. The electrons travel in the negative x-direction. Therefore, the force, F_y is in the positive y-direction due to the negative charge and the electrons move to the right, just like holes. The balancing electric field, ε_y, now has the opposite sign, which results in a negative Hall voltage.

To calculate the Hall field, we first calculate the Lorentz force acting on the free carriers :

$$\vec{F} = q\left(\vec{\varepsilon} + \vec{v} + \vec{B}\right) \qquad \text{... (2.34)}$$

We now assume that the carriers can only flow along the x-direction and label their velocity vx. The Lorentz force then becomes :

$$\vec{F} = q\vec{\varepsilon} + q\begin{vmatrix} \vec{e}_x & \vec{e}_y & \vec{e}_z \\ v_x & 0 & 0 \\ 0 & 0 & B_z \end{vmatrix} = q\varepsilon_x\,\vec{e}_x + q\left(\varepsilon_y - v_x\,B_z\right)\vec{e}_y + q\varepsilon_x\,\vec{e}_z \qquad \text{... (2.35)}$$

Since the carriers only flow along the x-direction, the net force must be zero along the y and z direction. As a result, the electric field is zero along the z direction and :

$$F_r = q\,(\varepsilon_r - v_x\,B_z) = 0 \qquad \text{... (2.36)}$$

which provides a relation between the electric field along the y-direction and the applied magnetic field. Equation (2.36) can also be rewritten as a function of the current density, J_x, using (2.2(b)), yielding :

$$\varepsilon_y = v_x\,B_z = \frac{J_x}{q p_{p0}}\,B_z \qquad \text{... (2.37)}$$

This electric field is called the Hall field. The Hall coefficient, R_H, is defined as the Hall field divided by the current density, J_x, and magnetic field, B_z :

$$R_H \overset{\Delta}{=} \frac{\varepsilon_r}{J_x\,B_z} = \frac{1}{q p_{p0}} \qquad \text{... (2.38)}$$

Once the Hall coefficient is obtained one easily finds the hole density :

$$p_{p0} = \frac{1}{qR_H} = \frac{J_x\,B_z}{q\varepsilon_r} \qquad \text{... (2.39)}$$

The carrier mobility can also be extracted from the Hall coefficient :

$$\mu_p = \frac{J_x}{q p_{p0}\,\varepsilon_r} = \frac{R_H}{\rho} \qquad \text{... (2.40)}$$

Where ρ is the resistivity of the semiconductor.

Alternatively, one can calculate the Hall coefficient from the measured current, I_x, and measured voltage, V_H :

$$R_H = \frac{V_H}{I_x\,B_z}\,\frac{tL}{W} \qquad \text{... (2.41)}$$

A measurement of the Hall voltage is often used to determine the type of semiconductor (n-type or p-type) the free carrier density and the carrier mobility. Repeating the measurement at different temperatures allows one to measure the free carrier density as well as the mobility as a function of temperature. Since the measurement can be done on a small piece of uniformly doped material it is by far the easiest measurement to determine the carrier mobility. It should be noted that the scattering mechanisms in the presence of a magnetic field are different and that the measured Hall mobility can differ somewhat from the drift mobility. A measurement of the carrier density versus temperature provides information regarding the ionization energies of the donors and acceptor that are present in the semiconductor. While the interpretation of the Hall measurement is straightforward in the case of a single dopant, multiple types of impurities and the presence of electrons and holes can make the interpretation non-trivial.

2.9 CARRIER RECOMBINATION AND GENERATION

Recombination of electrons and holes is a process by which both carriers annihilate each other : electrons occupy—through one or multiple steps—the empty state associated with a hole. Both carriers eventually disappear in the process. The energy difference between the initial and final state of the electron is released in the process. This leads to one possible classification of the recombination processes. In the case of radiative recombination, this energy is emitted in the form of a photon. In the case of non-radiative recombination, it is passed on to one or more phonons and in the case of Auger recombination it is given off in the form of kinetic energy to another electron. Another classification scheme considers the individual energy levels and particles involved. These different processes are further illustrated with Fig. 2.9.

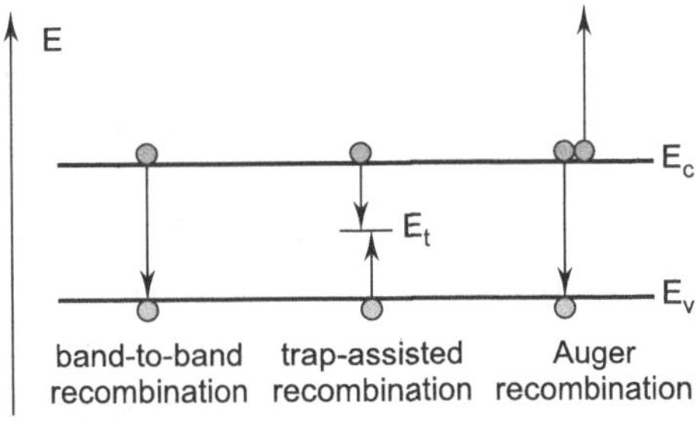

Fig. 2.9 *Carrier recombination in semiconductors*

Band-to-band recombination occurs when an electron moves from its conduction band state into the empty valence band state associated with the hole. This band-to-band transition is typically also a radiative transition in direct bandgap semiconductors.

Trap-assisted recombination occurs when an electron falls into a "trap", an energy level within the bandgap caused by the presence of a foreign atom or a structural defect. Once the trap is filled it cannot accept another electron. The electron occupying the trap, in a second step, moves into an empty valence band state, thereby completing the recombination process. One can envision this process as a two-step transition of an electron from the conduction band to the valence band or as the annihilation of the electron and hole, which meet each other in the trap. We will refer to this process as Shockley-Read-Hall (SRH) recombination.

Auger recombination is a process in which an electron and a hole recombine in a band-to-band transition, but now the resulting energy is given off to another electron or hole. The involvement of a third particle affects the recombination rate so that we need to treat Auger recombination differently from band-to-band recombination.

Each of these recombination mechanisms can be reversed leading to carrier generation rather than recombination. A single expression will be used to describe recombination as well as generation for each of the above mechanisms.

In addition, there are generation mechanisms, which do not have an associated recombination mechanism, such as generation of carriers by light absorption or by a high-energy electron/particle beam. These processes are referred to as ionization processes. Impact ionization, which is the generation mechanism associated with Auger recombination, also belongs to this category. The generation mechanisms are illustrated with Fig. 2.10.

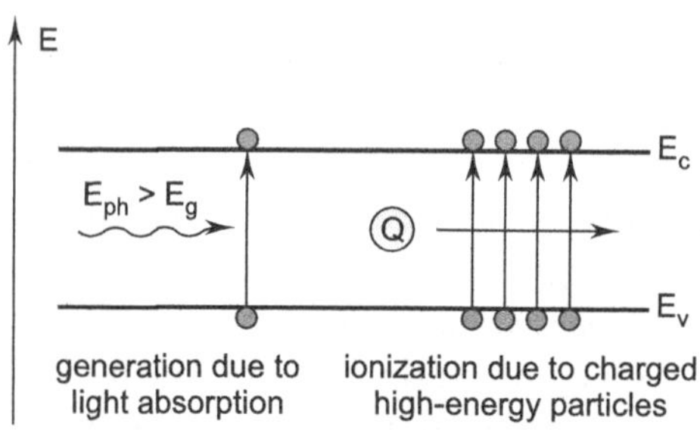

Fig. 2.10 *Carrier generation due to light absorption and ionization due to high-energy particle beams*

Carrier generation due to light absorption occurs if the photon energy is large enough to raise an electron from the valence band into an empty conduction band state, thereby generating one electron-hole pair. The photon energy needs to be larger than the bandgap energy to satisfy this condition. The photon is absorbed in this process and the excess energy, $E_{ph} - E_g$, is added to the electron and the hole in the form of kinetic energy.

Carrier generation or ionization due to a high-energy beam consisting of *charged* particles is similar except that the available energy can be much larger than the bandgap energy so that multiple electron-hole pairs can be formed. The high-energy particle gradually loses its energy and eventually stops. This generation mechanism is used in semiconductor-based nuclear particle counters. As the number of ionized electron-hole pairs varies with the energy of the particle, one can also use such detector to measure the particle energy.

Finally, there is a generation process called *impact ionization,* the generation mechanism that is the counterpart of Auger recombination. Impact ionization is caused by an electron/hole with an energy, which is much larger/smaller than the conduction/ valence band edge. The detailed mechanism is illustrated with Fig. 2.11.

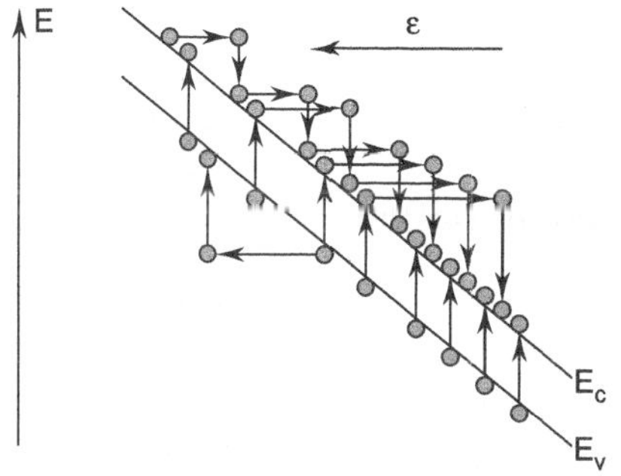

Fig. 2.11 *Impact ionization and avalanche multiplication of electrons and holes in the presence of a large electric field*

The excess energy is given off to generate an electron-hole pair through a band-to-band transition. This generation process causes avalanche multiplication in semiconductor diodes under high reverse bias : As one carrier accelerates in the electric field it gains energy. The kinetic energy is given off to an electron in the valence band, thereby creating an electron-hole pair. The resulting two electrons can create two more electrons which generate four more causing an avalanche multiplication effect. Electrons as well as holes.

2.9.1 Simple Recombination-Generation Model

A simple model for the recombination-generation mechanisms states that the recombination-generation rate is proportional to the excess carrier density. It acknowledges the fact that no net recombination takes place if the carrier density equals the thermal equilibrium value. The resulting expression for the recombination of electrons in a *p*-type semiconductor is given by :

$$U_n = R_n - G_n = \frac{n_p - n_{p0}}{\tau_n} \quad \text{... (2.42)}$$

and similarly for holes in an *n*-type semiconductor :

$$U_p = R_p - G_p = \frac{p_n - p_{n0}}{\tau_p} \quad \text{... (2.43)}$$

where the parameter τ can be interpreted as the average time after which an excess minority carrier recombines.

We will show for each of the different recombination mechanisms that the recombination rate can be simplified to this form when applied to minority carriers in a "quasi-neutral" semiconductor. The above expressions are therefore only valid under these conditions. The recombination rates of the majority carriers equals that of the minority carriers since in steady state recombination involves an equal number of holes and electrons. Therefore, the recombination rate of the **majority** carriers depends on the excess-**minority**-carrier-density as the minority carriers limit the recombination rate.

Recombination in a depletion region and in situations where the hole and electron density are close to each other cannot be described with the simple model and the more elaborate expressions for the individual recombination mechanisms must be used.

2.9.2 Band-to-Band Recombination

Band-to-band recombination depends on the density of available electrons and holes. Both carrier types need to be available in the recombination process. Therefore, the rate is expected to be proportional to the product of n and p. Also, in thermal equilibrium, the recombination rate must equal the generation rate since there is no net recombination or generation. As the product of n and p equals n_i^2 in thermal equilibrium, the net recombination rate can be expressed as :

$$U_{b-b} = b\left(np - n_i^2\right) \qquad \text{... (2.44)}$$

where b is the bimolecular recombination constant.

2.9.3 Trap Assisted Recombination

The net recombination rate for trap-assisted recombination is given by :

$$U_{SHR} = \frac{pn - n_i^2}{p + n + 2n_i \cosh\left(\dfrac{E_i - E_t}{kT}\right)} N_t v_{th} \sigma \qquad \text{... (2.45)}$$

The derivation of this equation is beyond the scope of this text.

This expression can be further simplified for $p >> n$ to :

$$U_n = R_n - G_n = \frac{n_p - n_{p0}}{\tau_n} \qquad \text{... (2.46)}$$

and for $n >> p$ to :

$$U_p = R_p - G_p = \frac{p_n - p_{n0}}{\tau_p} \qquad \text{... (2.47)}$$

were $$\tau_n = \tau_p = \frac{1}{N_t v_{th} \sigma} \qquad \text{... (2.48)}$$

2.9.4 Surface Recombination

Recombination at surfaces and interfaces can have a significant impact on the behavior of semiconductor devices. This is because surfaces and interfaces typically contain a large number of recombination centres because of the abrupt termination of the semiconductor crystal, which leaves a large number of electrically active states. In addition, the surfaces and interfaces are more likely to contain impurities since they are exposed during the device fabrication process. The net recombination rate due to trap-assisted recombination and generation is given by :

$$U_{s,\mathrm{SHR}} = \frac{pn - n_i^2}{p + n + 2n_i \cosh\left(\frac{E_i - E_{st}}{kT}\right)} N_{st}\, v_{th}\, \sigma_s \qquad \text{... (2.49)}$$

This expression is almost identical to that of Shockley-Hall-Read recombination. The only difference is that the recombination is due to a two-dimensional density of traps, N_{ts}, as the traps only exist at the surface or interface.

This equation can be further simplified for minority carriers in a quasi-neutral region. For instance for electrons in a quasi-neutral *p*-type region $p >> n$ and $p >> n_i$ so that for $E_i = E_{st}$, it can be simplified to :

$$U_{s,n} = R_{s,n} - G_{s,n} = v_s\,(n_p - n_{p0}) \qquad \text{... (2.50)}$$

where the recombination velocity v_s is given by,

$$v_s = N_{st}\, v_{\mathrm{th}}\, \sigma_s$$

2.9.5 Auger Recombination

Auger recombination involves three particles : An electron and a hole, which recombine in a band-to-band transition and give off the resulting energy to another electron or hole. The expression for the net recombination rate is therefore similar to that of band-to-band recombination but includes the density of the electrons or holes, which receive the released energy from the electron-hole annihilation :

$$U_{\mathrm{Auger}} = \Gamma_n\, n\left(np - n_i^2\right) + \Gamma_p\, p\left(np - n_i^2\right) \qquad \text{... (2.51)}$$

The two terms correspond to the two possible mechanisms.

2.9.6 Generation Due to Light

Carriers can be generated in semiconductors by illuminating the semiconductor with light. The energy of the incoming photons is used to bring an electron from a lower energy level to a higher energy level. In the case where an electron is removed from the valence band and added to the conduction band, an electron-hole pair is generated. A necessary condition is that the energy of the photon, E_{ph}, is larger than the bandgap energy, E_g. As the energy of the photon is given off to the electron, the photon no longer exists.

If each absorbed photon creates one electron-hole pair, the electron and hole generation rates are given by :

$$G_{\mathrm{p,light}} = G_{n,\mathrm{light}} = \alpha \frac{P_{opt}(x)}{E_{ph} A} \qquad \text{... (2.52)}$$

where α is the absorption coefficient of the material at the energy of the incoming photon. The absorption of light in a semiconductor causes the optical power to decrease with distance. This effect is described mathematically by :

$$\frac{dP_{opt}(x)}{dx} = -a p_{\mathrm{opt}}(x) \qquad \text{... (2.53)}$$

The calculation of the generation rate of carriers therefore requires first a calculation of the optical power within the structure from which the generation rate can then be obtained using (2.52).

Example. *Calculate the electron and hole densities in an n-type silicon wafer ($N_d = 10^{17}$ cm^{-3}) illuminated uniformly with 10 mW/cm^2 of red light (E_{ph} = 1.8 eV). The absorption coefficient of red light in silicon is 10^{-3} cm^{-1}. The minority carrier lifetime is 10 μs.*

Solution. The generation rate of electrons and holes equals :

$$G_n = G_p = \alpha \frac{P_{opt}}{E_{ph} A} = 10^{-3} \frac{10^{-2}}{1.8 \times 1.6 \times 10^{-19}} = 3.5 \times 10^{13} \text{ cm}^{-3} \text{ s}^{-1}$$

where the photon energy was converted into Joules. The excess carrier densities are then obtained from :

$$\delta n = \delta p = \tau_p G_p = 10 \times 10^{-6} \times 3.5 \times 10^{13} = 3.5 \times 10^{8} \text{ cm}^{-3} \text{ s}^{-1}$$

The excess carrier densities are then obtained from : So that the electron and hole densities equal:

$$n = n_o + \delta n = 10^{17} + 3.5 \times 10^{13} = 10^{17} \text{ cm}^{-3} \text{ s}^{-1}$$

2.9.7 Energy Loss in Recombination Processes

The energy loss in a recombination process equals the difference between the electron and hole quasi-Fermi energies as the energy loss is only due to the energy of the particles, which are lost :

$$\Delta U = F_n - F_p$$

No heat or work is removed from the system, just the energy associated with the particles. The energy lost in the recombination process can be converted in heat or light depending on the details of the process.

2.9.8 Minority Carrier Life Time

During the recombination process one hole and one electron disappear and the time taken for this process is called the minority carrier life time.

Under the thermal equilibrium, the equilibrium densities of electrons n_0 and holes p_0 remains constant though the generation and recombination of carriers are constantly taking place.

If G is the equilibrium generation rate and R the recombination rate then

$$R = G$$

or $$R - G = 0$$

In the case of band-to-band recombination, the recombination rate is linearly proportional to the number of electrons and holes present.

i.e., $$R \propto n_0 p_0$$

or $$R = r n_0 p_0 \qquad r = \text{proportionality constant.}$$

so, $$r n_0 p_0 - G = 0.$$

Let, Δn and Δp is the excess of holes and electrons generated.

Here, $\Delta n = \Delta p$.

So, net rate of excess hole-carrier recombination at any time is given by :

$$-\frac{d}{dt}(\Delta p) = r\,(n_0 + \Delta n)\,(p_0 + \Delta p) - r\,n_0\,p_0$$

For, small introducing of excess carriers the density of excess carrier concentration is very small, for this reason r has not changed.

So, $\frac{d}{dt}(\Delta p) = -\,r\,(n_0 + p_0)\,\Delta\phi$; neglecting the 2nd order terms. If we consider or n-type material ($n_0 >> p_0$) under low-level injection ($\Delta p << n_0$), then $\frac{d(\Delta p)}{dt} = -\,r\,n_0\,(\Delta p)$.

The solution to the equation is an exponential delay from the initial excess concentration.

So, $\Delta p\,(t) = \Delta p\,(0)\,e^{-r n_0 t} = \Delta p\,(0)\,e^{-t/\tau_{p0}}$. Where,

$$\tau_{p0} = (r\,n_0)^{-1}$$

excess minority carrier lifetime of holes in n-type material.

i.e., $\frac{d}{\Delta p}(\Delta p) = -\frac{dt}{\tau}$

Integrating, we get,

$$\Delta p = A\,e^{-t/\tau}$$

Similarly, $\tau_{n0} = (r\,p_0)^{-1}$.

Let, $t = 0, \Delta p = (\Delta p)_0$

Then, $A = (\Delta p)_0$

$\therefore$ $\Delta p = (\Delta p)_0\,e^{-t/\tau}$

After, $\tau = r$, then,

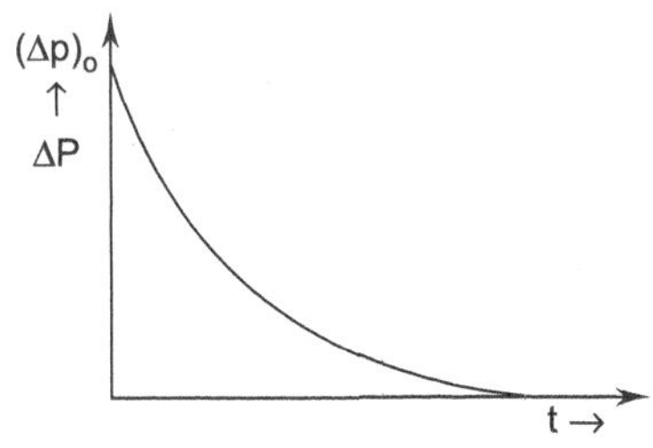

Fig. 2.12 *Variation of excess charge carriers with time*

$$\boxed{\frac{\Delta p}{(\Delta p)_0} = \frac{1}{e}}$$

* For the intrinsic semiconductor,

$$\tau = \tau_{p0} + \tau_{n0}.$$

Differently says that, the minority carrier life time can be defined as the time taken for the excess charge carriers to reduce to $\frac{1}{e}$ times its initial, value, once the source generating there excess charge carriers is cut off.

Recombination rate of the majority carrier electrons will be the same as that of the minority carrier holes. So, that

$$\boxed{R_n = R_p = \frac{\Delta n}{\tau_{p0}}.}$$

2.10 CONTINUITY EQUATION

When a semiconductor bar is illuminated on its one face, then excess carriers are generated at the exposed surface. These charge carriers diffuse throughout the material.

Hence, the carrier concentration in the body of the sample is a function of both time and distance.

2.10.1 Derivation

The continuity equation describes a basic concept, namely that a change in carrier density overtime is due to the difference between the incoming and outgoing flux of carriers plus the generation and minus the recombination. The flow of carriers and recombination and generation rates are illustrated with Fig. 2.13.

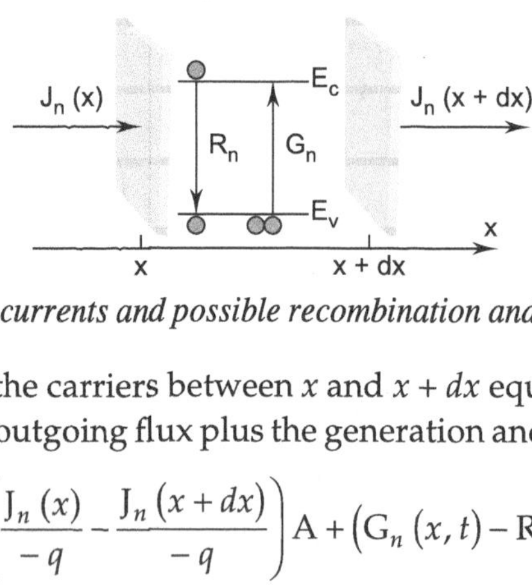

Fig. 2.13 *Electron currents and possible recombination and generation processes*

The rate of change of the carriers between x and $x + dx$ equals the difference between the incoming flux and the outgoing flux plus the generation and minus the recombination:

$$\frac{\partial n(x,t)}{\partial t} A\,dx = \left(\frac{J_n(x)}{-q} - \frac{J_n(x+dx)}{-q}\right) A + \left(G_n(x,t) - R_n(x,t)\right) A\,dx \qquad \text{... (2.54)}$$

where $n(x, t)$ is the carrier density, A is the area, $G_n(x, t)$ is the generation rate and $R_n(x, t)$ is the recombination rate. Using a Taylor series expansion,

$$J_n(x+dx) = J_n(x) + \frac{dJ_n(x)}{dx}dx \qquad \text{... (2.55)}$$

This equation can be formulated as a function of the derivative of the current :

$$\frac{\partial n(x,t)}{\partial t} = \frac{1}{q}\frac{\partial J_n(x,t)}{\partial x} + G_n(x,t) - R_n(x,t) \qquad \text{... (2.56)}$$

δ and similarly for holes one finds :

$$\frac{\partial p(x,t)}{\partial t} = -\frac{1}{q}\frac{\partial J_p(x,t)}{\partial x} + G_p(x,t) - R_p(x,t) \quad \text{... (2.57)}$$

A solution to these equations can be obtained by substituting the expression for the electron and hole current, (2.31) and (2.32). This then yields two partial differential equations as a function of the electron density, the hole density and the electric field. The electric field itself is obtained from Gauss's law.

$$\frac{\partial n(x,t)}{\partial t} = \mu_n n\frac{\partial \varepsilon(x,t)}{\partial x} + \mu_n \varepsilon\frac{\partial n(x,t)}{\partial x} + D_n\frac{\partial^2 n(x,t)}{\partial x^2} + G_n(x,t) - R_n(x,t) \quad \text{... (2.58)}$$

$$\frac{\partial p(x,t)}{\partial t} = -\mu_p p\frac{\partial \varepsilon(x,t)}{\partial x} - \mu_p \varepsilon\frac{\partial p(x,t)}{\partial x} + D_p\frac{\partial^2 p(x,t)}{\partial x^2} + G_p(x,t) - R_p(x,t) \quad \text{... (2.59)}$$

2.10.2 The Diffusion Equation

In the quasi-neutral region—a region containing mobile carriers, where the electric field is small—the current is due to diffusion only. In addition, we can use the simple recombination model for the net recombination rate since the recombination rates depend only on the minority carrier density. This leads to the time-dependent diffusion equations for electrons in *p*-type material and for holes in *n*-type material :

$$\frac{\partial n(x,t)}{\partial t} = D_n\frac{\partial^2 n_p(x,t)}{\partial x^2} - \frac{n_p(x,t) - n_{p0}}{\tau_n} \quad \text{... (2.60)}$$

$$\frac{\partial p(x,t)}{\partial t} = D_p\frac{\partial^2 p_n(x,t)}{\partial x^2} - \frac{p_n(x,t) - p_{n0}}{\tau_p} \quad \text{... (2.61)}$$

2.10.3 Steady State Solution to the Diffusion Equation

In steady state, the partial derivatives with respect to time are zero, yielding :

$$0 = D_n\frac{d^2 n_p(x)}{dx^2} - \frac{n_p(x) - n_{p0}}{\tau_n} \quad \text{... (2.62)}$$

$$0 = D_p\frac{d^2 p_n(x)}{dx^2} - \frac{p_n(x) - p_{n0}}{\tau_p} \quad \text{... (2.63)}$$

The general solution to these second order differential equations are :

$$n_p(x \le -x_p) = n_{p0} + Ce^{-(x+x_p)/L_n} + De^{(x+x_p)/L_n} \quad \text{... (2.64)}$$

$$p_n(x \ge x_n) = p_{n0} + Ae^{-(x-x_n)/L_p} + Be^{(x-x_n)/L_p} \quad \text{... (2.65)}$$

where L_n and L_p are the diffusion lengths given by :

$$L_n = \sqrt{D_n \tau_n} \quad \text{... (2.66)}$$

$$L_p = \sqrt{D_p \tau_p} \quad \text{... (2.67)}$$

The diffusion constants, D_n and D_p, are obtained using the Einstein relations (2.29) and (2.30). The diffusion equations can also be written as a function of the *excess* >/i> *carrier densities, ∂n and ∂p, which are related to the total carrier densities, n and p, and the thermal equilibrium densities, n_0 and p_0, by :*

$$n = n_0 + \delta n \quad \text{... (2.68)}$$

$$p = p_0 + \delta p \quad \text{... (2.69)}$$

yielding :

$$0 = \frac{d^2(\delta n_p)}{dx^2} - \frac{\delta n_p}{L_n^2} \quad \text{... (2.70)}$$

$$0 = \frac{d^2(\delta p_n)}{dx^2} - \frac{\delta p_n}{L_p^2} \quad \text{... (2.71)}$$

The diffusion equation will be used to calculate the diffusion current in p-n junctions and bipolar transistors.

2.11 SURFACE EFFECTS

(A) **Surface States :** When a semiconductor is abruptly terminated, the perfect periodic nature of the ideal single-crystal lattice ends abruptly at the surface. The disorders of the periodic potential function forms the allowed electronic energy states within the forbidden energy gap, because we know that simple defects in the semiconductor would create discrete energy states within the bandgap.

The abrupt termination of the periodic potential at the surface results in a distribution of allowed energy states within the bandgap, (shown in Fig. 2.14) along with the discrete energy states in the bulk semiconductor.

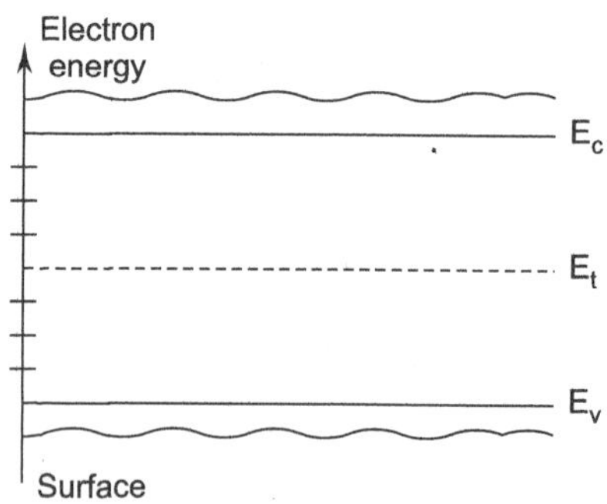

Fig. 2.14 *Distribution of surface states within the forbidden bandgap*

The Shockley-Read-Hall recombination theory shows that the excess minority carrier lifetime is inversely proportional to the density of trap states.

Again, the density of traps at the surface is larger than in the bulk. For this reason, the excess minority carrier lifetime at the surface will be smaller than the corresponding lifetime in the bulk material.

For example, for an extrinsic *n*-type semiconductor, the recombination rate of excess carriers in the bulk,

$$R - \frac{\delta p}{\tau_{p0}} = \frac{\delta p_B}{\tau_{p0}}$$

where δp_B is the concentration of excess minority carrier holes in the bulk material.

The recombination rate of excess carriers at the surface as,

$$R_s = \frac{\delta p_s}{\tau_{pos}}.$$

where δ_{ps} = excess minority carrier hole concentration at the surface.

$\tau_{pos} \Rightarrow$ the excess minority carrier hole life time at the surface.

Assume that excess carriers are being generated at a constant rate throughout the entire semiconductor material.

In steady state, the generation rate is equal to the recombination rate for the case of a homogeneous, infinite semiconductor.

Using this argument, the recombination rates at the surface (R_s) and in the bulk material (R_B) must be equal.

Since $\tau_{p0s} < \tau_{p0}$, then the excess carrier (minority) concentration at the surface is smaller than the excess minority carrier concentration in the bulk region.

or $\delta p_S > \delta p_B$

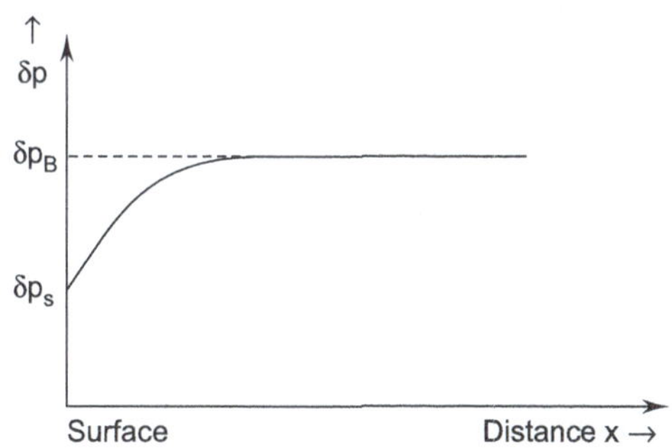

Fig. 2.15 *Steady state excess hole concentration versus distance from a semiconductor surface*

(B) **Surface Recombination Velocity :** A gradient in the excess carrier concentration exists near the surface as shown in Fig. 5, excess carriers from the bulk region diffuse towards the surface where they recombine. This diffusion towards the surface can be described by the equation,

$$-D_p \left[\hat{n} \cdot \frac{d(\delta p)}{dx} \right] \Bigg|_{surf} = s\,\delta p \,\Big|_{surf}$$

where each side of the equation is evaluated at the surface.

The parameter $\hat{n}$ is the unit towards vector normal to the surface.

From the Fig. 5, $\frac{d(\delta p)}{dx}$ is positive quantity and $\hat{n}$ is negative quantity, so that the parameter s is positive quantity. The units of $s \Rightarrow$ cm/sec or velocity. The parameter s is called surface recombination velocity.

If the excess concentration at the surface is equal to the excess concentration at the bulk, then the gradient term would be zero and the surface recombination velocity would be zero.

As the excess concentration at the surface becomes smaller, the gradient term becomes larger and the surface recombination velocity increases.

Sin $\hat{n} = -1$ then,

$$D_p \left.\frac{d(\delta p)}{dx}\right|_{surf} = s\,\delta p\,|_{surf}$$

The excess minority carrier hole concentration with the help of surface recombination velocity, can be written as,

$$\delta p(x) = g\,\tau_{p0}\left(1 - \frac{s L_p\, e^{-x/L_p}}{D_p + s L_p}\right)$$

where, L_p = Diffusion length

$$= \sqrt{D_p\,\tau_{p0}}\,.$$

2.12 CHARGE STORAGE AND DIODE TRANSIENTS

The p-n junction is mainly used as an electrical switch.

In forward bias, it acts as the 'on' state, a relatively large current can be produced by a small applied voltage.

In reverse bias, it acts as the 'off' state, only a very small current will exist.

To know the speed of the p-n junction diode in switching state, we will simply state the equations that describe the switching times without any mathematical derivations.

(A) **The Turn-off Transient :** Let we want to switch a diode from the forward bias on state to the reverse-bias off state.

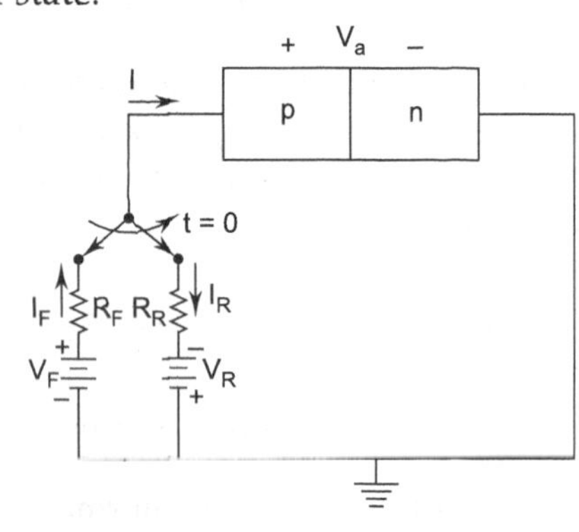

Fig. 2.16 *For $t < 0$, the forward current is,* $I = I_F = \dfrac{V_F - V_a}{R_F}$

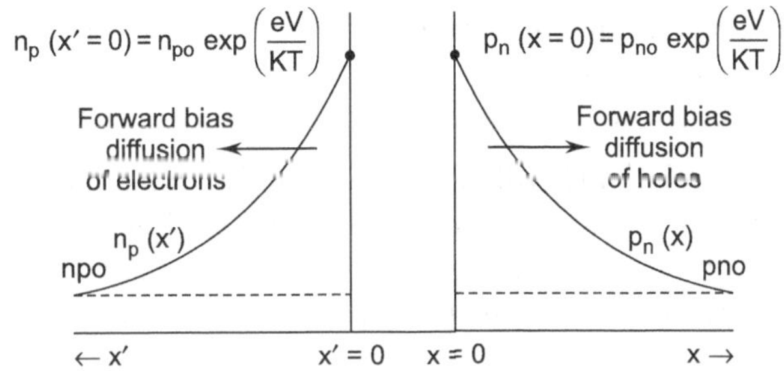

Fig. 2.17 *Steady-state forward-bias minority carrier concentrations*

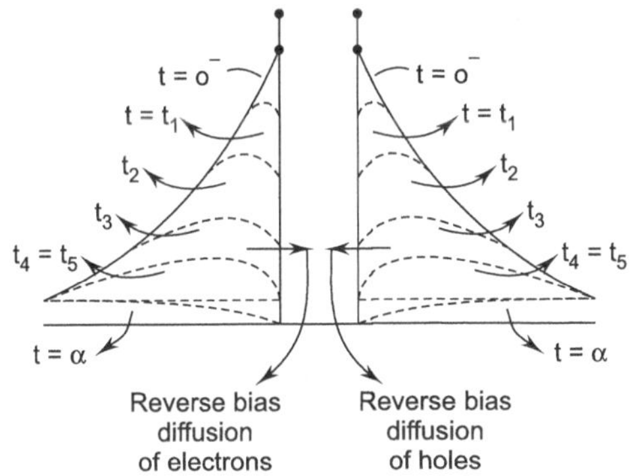

Fig. 2.18 *Minority carrier concentrations at various times during switching*

The excess minority carrier charge stored in both the *p* and *n* regions of the diode. The excess minority carrier concentrations at the space charge edges are supported by the forward-bias junction voltage V_a.

When the voltage is switched from the forward—to the reverse bias state, the excess minority carrier concentrations at the space charge edges can no longer be supported and they start to decrease, as shown in Fig. 2.18.

The collapse of the minority carrier concentrations at the edges of the space charge region leads to large concentration gradients and diffusion currents in the reverse-bias direction.

If, the voltage across the diode junction to small compared with V_R, then the reverse bias current is limited to approximately :

$$J = -J_R \approx \frac{-V_R}{R_R}$$

The junction capacitances do not allow the junction voltage to change instantaneously. If the current I_R were larger than this value, there would be a forward-bias voltage across the junction, which would violate our assumption of a reverse-bias current.

If the current I_R were smaller than this value, there, would be a reverse-bias voltage across the junction, which means that the junction voltage would have changed instantaneously.

Since the reverse current is limited to the value given by equation above, the reverse bias density gradient is constant ; thus, the minority carrier concentrations at the space charge edge decrease with time.

This reverse current I_R will be approximately constant for $0+ \leq t \leq t_s$, where t_s is called the storage time. The storage time is the length of time required for the minority carrier concentrations at the space charge edge to reach the thermal equilibrium values. After this time, the voltage across the junction will begin to change. The current characteristic is shown in Fig. 2.19.

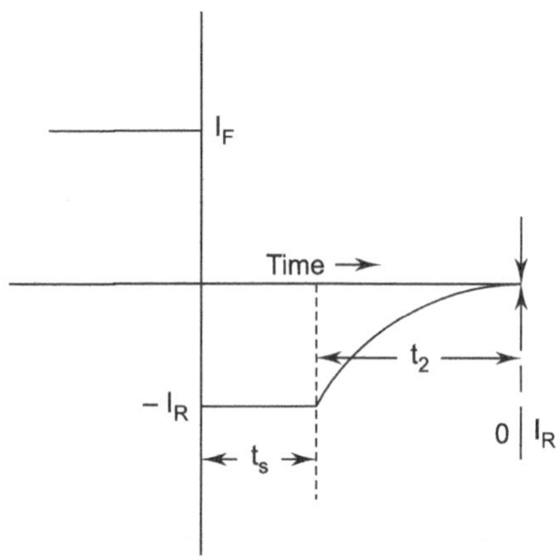

Fig. 2.19 *Current characteristic versus time during diode switching.*

The reverse current is the flow of the stored minority carrier charge, which is the difference between the minority carrier concentrations at $t = 0^-$ and $t = \alpha$ as shown in Fig. 8.

The storage time t_s can be determined by solving the time—dependent continuity equation. If we consider as one-sided $p^+ n$ junction, the storage time is determined from the equation.

$$\text{erf}\sqrt{\frac{t_s}{\tau_{p0}}} = \frac{I_F}{I_F + I_R}$$

where erf (x) is known as the error function. An approximate solution for the storage time can be obtained as :

$$t_s \approx \tau_{p0} \ln\left(1 + \frac{I_F}{I_R}\right)$$

The recovery phase for $t > t_s$ is the time required for the junction to reach its steady state reverse-bias condition. The remainder of the excess charge is being removed and the space charge width is increasing to the reverse-bias value.

The delay time t_2 is determined from

$$\operatorname{erf}\sqrt{\frac{t_2}{\tau_{p0}}} + \frac{\exp\left(\frac{-t_2}{\tau_{p0}}\right)}{\sqrt{\frac{\pi t_2}{\tau_{p0}}}} = 1 + 0.1\left(\frac{I_R}{I_F}\right)$$

The total turn-off time is the sum of t_s and t_2.

To switch the diode quickly, we need to be able to produce a large reverse current as well as have a small minority carrier lifetime. In the design of diode circuits, then, the designer must provide a path for the tramient reverse-bias current pulse in order to be able to switch the diode quickly.

QUESTIONS

- Name the two transport mechanisms in semiconductors.
- Describe the microscopic behavior of electrons and holes in a semiconductor.
- Define the mobility.
- Explain why the mobility in a semiconductor depends on the doping density.
- Define the resistivity and conductivity of a semiconductor.
- Explain why the velocity in a semiconductor is limited.
- What is the driving force, which causes diffusion ?
- Explain the relation between the mean free path, the scattering time and the thermal velocity.
- List three recombination-generation mechanisms.
- Explain why the net recombination rate as described by the simple model depends on the excess carrier density.
- Describe the continuity equation in words.
- What is the diffusion length and how does it relate to the diffusion constant and the minority carrier lifetime ?
- What is the drift-diffusion model ?

3

Metal-Semiconductor Junctions

3.1 INTRODUCTION

Metal-to-semiconductor contacts are of great importance since they are present in every semiconductor device. They can behave either as a Schottky barrier or as an ohmic contact dependent on the characteristics of the interface. This chapter contains an analysis of the electrostatics of the M-S junction (*i.e.*, the charge, field and potential distribution within the device) followed by a derivation of the current voltage characteristics due to diffusion, thermionic emission and tunneling and a discussion of the non-ideal effects in Metal-Semiconductor junctions.

3.2 STRUCTURE AND PRINCIPLE OF OPERATION

3.2.1 Structure

The structure of a metal-semiconductor junction is shown in Fig. 3.1. It consists of a metal contacting a piece of semiconductor. An ideal Ohmic contact such that no potential exists between the metal and the semiconductor, is made to the other side of the semiconductor. The sign convention of the applied voltage and current is also shown on Fig. 3.1.

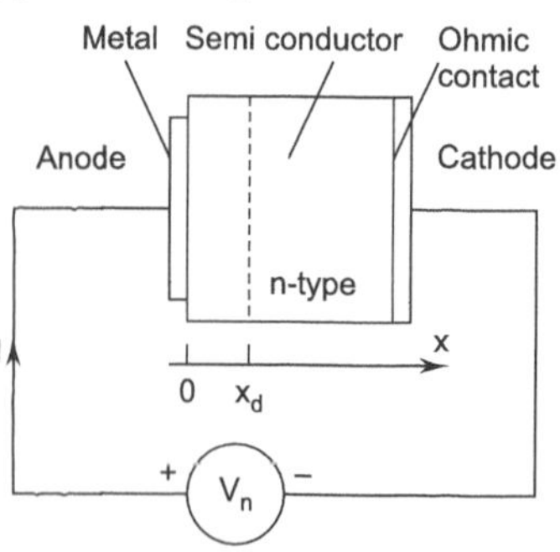

Fig. 3.1 *Structure and sign convention of a metal-semiconductor junction*

3.2.2 Flatband Diagram and Built-in Potential

The barrier between the metal and the semiconductor can be identified on an energy band diagram. To construct such diagram we first consider the energy band diagram of the metal and the semiconductor, and align them using the same vacuum level as shown in Fig. 3.2 (*a*). As the metal and semiconductor are brought together, the Fermi energies of the metal and the semiconductor do not change right away. This yields the flatband diagram of Fig. 3.2 (*b*).

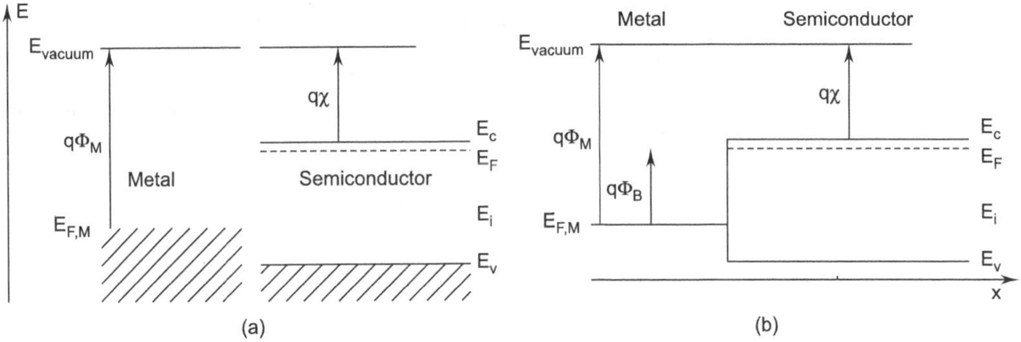

Fig. 3.2 *Energy band diagram of the metal and the semiconductor before (a) and after (b) contact is made*

The barrier height, ϕ_B, is defined as the potential difference between the Fermi energy of the metal and the band edge where the majority carriers reside. From Fig. 3.2 (*b*) one finds that for an n-type semiconductor the barrier height is obtained from :

$$\phi_B = \Phi_M - \chi, \text{ for an } n\text{-type semiconductor} \quad \text{... (3.1)}$$

Where Φ_M is the work function of the metal and χ is the electron affinity. The work function of selected metals as measured in vacuum can be found in Table 3.1. For *p*-type material, the barrier height is given by the difference between the valence band edge and the Fermi energy in the metal :

$$\phi_B = \frac{E_g}{q} + \chi - \Phi_M, \text{ for a } p\text{-type semiconductor} \quad \text{... (3.2)}$$

A metal-semiconductor junction will therefore form a barrier for electrons and holes if the Fermi energy of the metal as drawn on the flatband diagram is somewhere between the conduction and valence band edge.

In addition, we define the built-in potential, ϕ_I, as the difference between the Fermi energy of the metal and that of the semiconductor.

$$\phi_I = \Phi_M - \chi - \frac{E_c - E_{F,n}}{q}, n\text{-type} \quad \text{... (3.3)}$$

$$\phi_I = \chi + \frac{E_c - E_{F,p}}{q} - \Phi_M, p\text{-type} \quad \text{... (3.4)}$$

The measured barrier height for selected metal-semiconductor junctions is listed in Table 3.1. These experimental barrier heights often differ from the ones calculated using (3.1) or (3.2). This is due to the detailed behavior of the metal-semiconductor interface. The ideal metal-semiconductor theory assumes that both materials are infinitely pure, that there is no interaction between the two materials nor is there an interfacial layer. Chemical reactions between the metal and the semiconductor alter the barrier height as do interface states at the surface of the semiconductor and interfacial layers. Some general trends however can still be observed. As predicted by (3.1), the barrier height on *n*-type semiconductors increases for metals with a higher work function as can be verified for silicon. Gallium arsenide on the other hand is known to have a large density of surface states so that the barrier height becomes virtually independent of the metal. Furthermore, one finds the barrier heights reported in the literature to vary widely due to different surface cleaning procedures.

Table 3.1 Workfunction of selected metals and their measured barrier height on Ge, Si and GaAs.

	Ag	*Al*	*Au*	*Cr*	*Ni*	*Pt*	*W*
ϕ_M (in vacuum)	4.3	4.25	4.8	4.5	4.5	5.3	4.6
***n*-Ge**	0.54	0.48	0.59		0.49		0.48
***p*-Ge**	0.5		0.3				
***n*-Si**	0.78	0.72	0.8	0.61	0.61	0.9	0.67
***p*-Si**	0.54	0.58	0.34	0.5	0.51		0.45
***n*-GaAs**	0.88	0.8	0.9			0.84	0.8
***p*-GaAs**	0.63		0.42				

3.2.3 Thermal Equilibrium

The flatband diagram, shown in Fig. 3.2 (*b*), is not a thermal equilibrium diagram, since the Fermi energy in the metal differs from that in the semiconductor. Electrons in the n-type semiconductor can lower their energy by traversing the junction. As the electrons leave the semiconductor, a positive charge, due to the ionized donor atoms, stays behind. This charge creates a negative field and lowers the band edges of the semiconductor. Electrons flow into the metal until equilibrium is reached between the diffusion of electrons from the semiconductor into the metal and the drift of electrons caused by the field created by the ionized impurity atoms. This equilibrium is characterized by a constant Fermi energy throughout the structure.

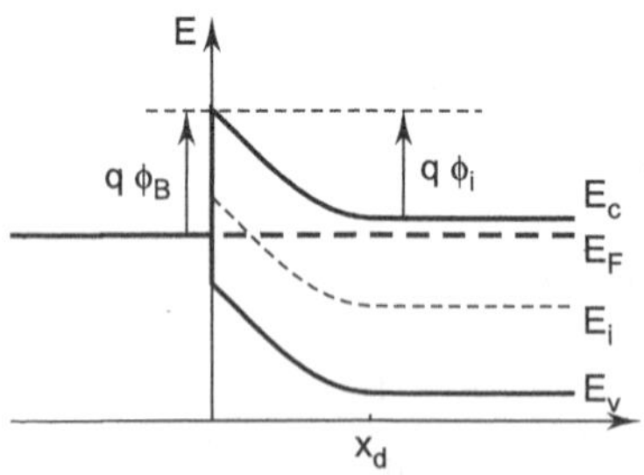

Fig. 3.3 *Energy band diagram of a metal-semiconductor contact in thermal equilibrium.*

It is of interest to note that in thermal equilibrium, *i.e.*, with no external voltage applied, there is a region in the semiconductor close to the junction ( ), which is depleted of mobile carriers. We call this the depletion region. The potential across the semiconductor equals the built-in potential, ϕ_i.

3.2.4 Forward and Reverse Bias

Operation of a metal-semiconductor junction under forward and reverse bias is illustrated with Fig. 3.4. As a positive bias is applied to the metal (Fig. 3.4 (*a*)), the Fermi energy of the metal is lowered with respect to the Fermi energy in the semiconductor. This results in a smaller potential drop across the semiconductor. The balance between diffusion and drift is disturbed and more electrons will diffuse towards the metal than the number drifting into the semiconductor. This leads to a positive current through the junction at a voltage comparable to the built-in potential.

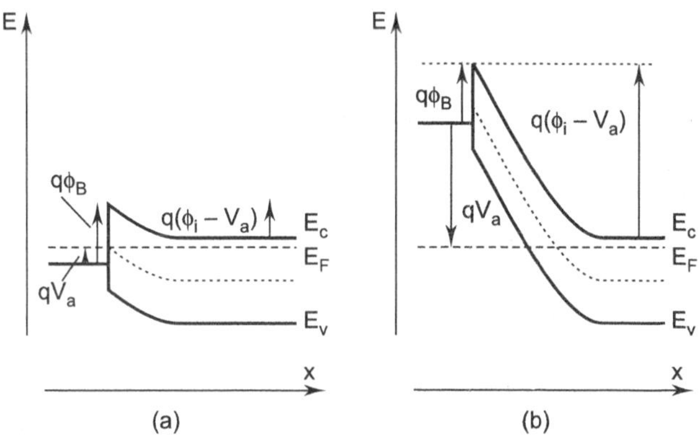

Fig. 3.4 *Energy band diagram of a metal-semiconductor junction under (a) forward and (b) reverse bias*

As a negative voltage is applied (Fig. 3.4 (*b*)), the Fermi energy of the metal is raised with respect to the Fermi energy in the semiconductor. The potential across the semiconductor now increases, yielding a larger depletion region and a larger electric field at the interface. The barrier, which restricts the electrons to the metal, is unchanged so that the flow of electrons is limited by that barrier independent of the applied voltage. The metal-semiconductor junction with positive barrier height has therefore a pronounced rectifying behavior. A large current exists under forward bias, while almost no current exists under reverse bias.

The potential across the semiconductor therefore equals the built-in potential, ϕ_i, minus the applied voltage, V_a.

$$\phi(x=\infty) - \phi(x=0) = \phi_i - V_a \quad \text{...(3.5)}$$

3.3 ELECTROSTATIC ANALYSIS

3.3.1 General Discussion – Poisson's Equation

The electrostatic analysis of a metal-semiconductor junction is of interest since it provides knowledge about the charge and field in the depletion region. It is also required to obtain the capacitance-voltage characteristics of the diode.

The general analysis starts by setting up Poisson's equation:

$$\frac{d^2\phi}{dx^2} = -\frac{\rho}{\varepsilon_s} = -\frac{q}{\varepsilon_s}\left(p - n + N_d^+ - N_a^-\right) \quad \text{... (3.6)}$$

where the charge density, ρ, is written as a function of the electron density, the hole density and the donor and acceptor densities. To solve the equation, we have to express the electron and hole density, n and p, as a function of the potential, ϕ, yielding :

$$\frac{d^2\phi}{dx^2} = \frac{2qn_i}{\varepsilon_s}\left(\sinh\frac{\phi - \phi_F}{V_t} + \sinh\frac{\phi_F}{V_t}\right) \quad \text{... (3.7)}$$

with

$$\sinh\frac{\phi_F}{V_t} = \frac{N_a^- - N_d^+}{2n_i} \quad \text{... (3.8)}$$

where the potential is chosen to be zero in the n-type region, where $x >> x_n$.

his second-order non-linear differential equation (3.7) can not be solved analytically. Instead we will make the simplifying assumption that the depletion region is fully depleted and that the adjacent neutral regions contain no charge. This full depletion approximation is the topic of section 3.3.2.

3.3.2 Full Depletion Approximation

The simple analytic model of the metal-semiconductor junction is base on the full depletion approximation. This approximation is obtained by assuming that the semiconductor is fully depleted over a distance x_d, called the depletion region. While this assumption does not provide an accurate charge distribution, it does provide very reasonable approximate expressions for the electric field and potential throughout the semiconductor.

3.3.3 Full Depletion Analysis

We now apply the full epletion approximation to an M-S junction containing an n-type semiconductor. We define the depletion region to be between the metal-semiconductor interface ($x = 0$) and the edge of the depletion region ($x = x_d$). The depletion layer width, x_d, is unknown at this point but will later be expressed as a function of the applied voltage.

To find the depletion layer width, we start with the charge density in the semiconductor and calculate the electric field and the potential across the semiconductor as a function of the depletion layer width. We then solve for the depletion layer width by requiring the potential across the semiconductor to equal the difference between the built-in potential and the applied voltage, $\phi_i - V_a$. The different steps of the analysis are illustrated by Fig. 3.5.

As the semiconductor is depleted of mobile carriers within the depletion region, the charge density in that region is due to the ionized donors. Outside the depletion region, the semiconductor is assumed neutral. This yields the following expressions for the charge density, ρ :

$$\rho(x) = qN_i \qquad\qquad 0 < x < x_d$$
$$\rho(x) = 0 \qquad\qquad x_d < x \qquad\qquad \text{... (3.9)}$$

where we assumed full ionization so that the ionized donor density equals the donor density, N_d. This charge density is shown in Fig. 3.5 (*a*). The charge in the semiconductor is exactly balanced by the charge in the metal, Q_M, so that no electric field exists except around the metal-semiconductor interface.

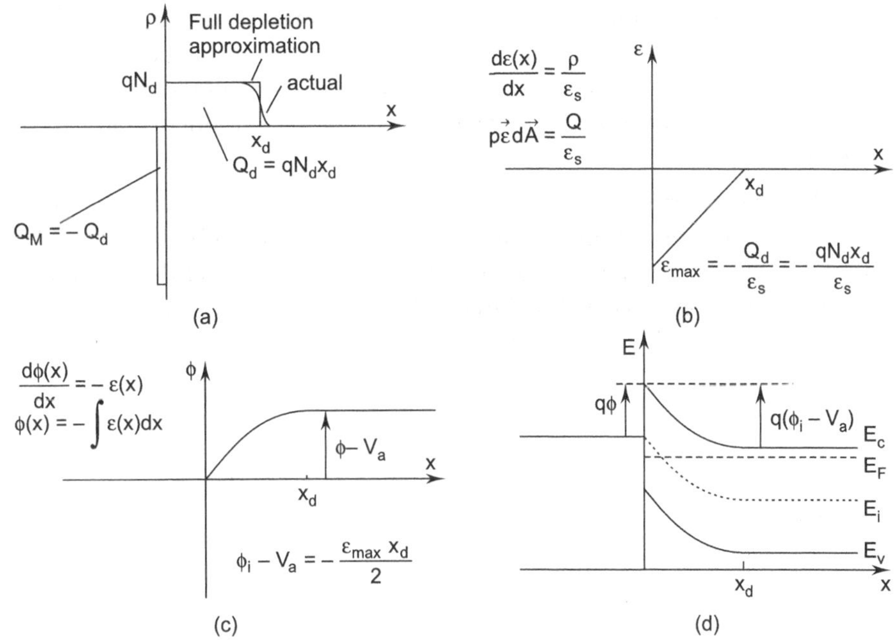

Fig. 3.5 *(a) Charge density, (b) electric field, (c) potential and (d) energy as obtained with the full depletion analysis*

Using Gauss's law we obtain electric field as a function of position, also shown in Fig. 3.5 (*b*) :

$$\varepsilon(x) = -\frac{qN_d}{\varepsilon_s}(x_d - x) \qquad\qquad 0 < x < x_d \qquad\qquad \text{... (3.10)}$$
$$\varepsilon(x) = 0 \qquad\qquad x_d \le x$$

where ε_s is the dielectric constant of the semiconductor. We also assumed that the electric field is zero outside the depletion region. It is expected to be zero there since a non-zero field would cause the mobile carriers to redistribute until there is no field. The depletion region does not contain mobile carriers so that there can be an electric field. The largest (absolute) value of the electric field is obtained at the interface and is given by :

$$\varepsilon(x = 0) = \frac{qN_d x_d}{\varepsilon_s} = -\frac{Q_d}{\varepsilon_s} \qquad\qquad \text{... (3.11)}$$

where the electric field was also related to the total charge (per unit area), Q_d, in the depletion layer. Since the electric field is minus the gradient of the potential, one obtains the potential by integrating the expression for the electric field, yielding :

$$\phi(x) = 0 \qquad x \le 0$$

$$\phi(x) = \frac{qN_d}{2\varepsilon_s}\left[x_d^2 - (x_d - x)^2\right] \qquad 0 < x < x_d \qquad \text{... (3.12)}$$

$$\phi(x) = \frac{qN_d x_d^2}{2\varepsilon_s} \qquad x_d \le x$$

We now assume that the potential across the metal can be neglected. Since the density of free carriers is very high in a metal, the thickness of the charge layer in the metal is very thin. Therefore, the potential across the metal is several orders of magnitude smaller that across the semiconductor, even though the total amount of charge is the same in both regions.

The total potential difference across the semiconductor equals the built-in potential, ϕ_i, in thermal equilibrium and is further reduced/increased by the applied voltage when a positive/negative voltage is applied to the metal. This boundary condition provides the following relation between the semiconductor potential at the surface, the applied voltage and the depletion layer width :

$$\phi_i - V_a = -\phi(x = 0) = \frac{qN_d x_d^2}{2\varepsilon_s} \qquad \text{... (3.13)}$$

Solving this expression for the depletion layer width, x_d, yields :

$$x_d = \sqrt{\frac{2\varepsilon_s(\phi_i - V_a)}{qN_d}} \qquad \text{... (3.14)}$$

3.4 JUNCTION CAPACITANCE

In addition, we can obtain the capacitance as a function of the applied voltage by taking the derivative of the charge with respect to the applied voltage yielding :

$$C_j = \left|\frac{dQ_d}{dV_a}\right| = \sqrt{\frac{q\varepsilon_s N_d}{2(\phi_i - V_a)}} = \frac{\varepsilon_s}{x_d} \qquad \text{... (3.15)}$$

The last term in the equation indicates that the expression of a parallel plate capacitor still applies. One can understand this once one realizes that the charge added/removed from the depletion layer as one decreases/increases the applied voltage is added/ removed only at the edge of the depletion region. While the parallel plate capacitor expression seems to imply that the capacitance is constant, the metal-semiconductor junction capacitance is not constant since the depletion layer width, x_d, varies with the applied voltage.

3.5 SCHOTTKY BARRIER LOWERING

Image charges build up in the metal electrode of a metal-semiconductor junction as carriers approach the metal-semiconductor interface. The potential associated with these charges reduces the effective barrier height. This barrier reduction tends to be rather small compared to the barrier height itself. Nevertheless this barrier reduction is of interest since it depends on the applied voltage and leads to a voltage dependence of the reverse bias current. Note that this barrier lowering is only experienced by a carrier while approaching the interface and will therefore not be noticeable in a capacitance-voltage measurement.

An energy band diagram of an n-type silicon Schottky barrier including the barrier lowering is shown in Fig. 3.6 :

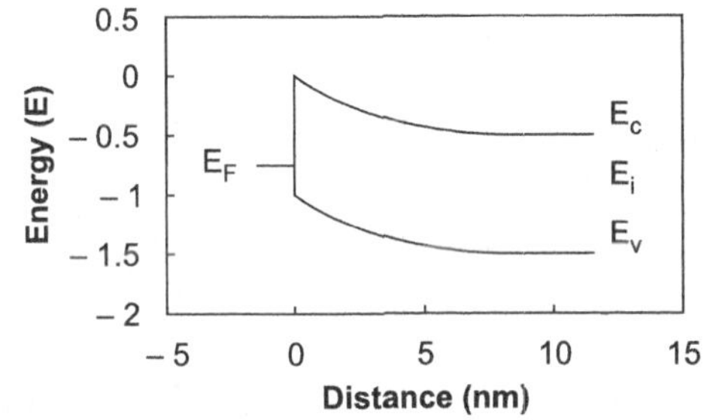

Fig. 3.6 *Energy band diagram of a silicon Schottky barrier with $\phi_B = 0.8$ V and $N_d = 10^{19}$ cm^3.*

Shown is the energy band diagram obtained using the full-depletion approximation, the potential reduction experienced by electrons, which approach the interface and the resulting conduction band edge. A rounding of the conduction band edge can be observed at the metal-semiconductor interface as well as a reduction of the height of the barrier.

The calculation of the barrier reduction assumes that the charge of an electron close to the metal-semiconductor interface attracts an opposite surface charge, which exactly balances the electron's charge so that the electric field surrounding the electron does not penetrate beyond this surface charge. The time to build-up the surface charge and the time to polarize the semiconductor around the moving electron is assumed to be much shorter than the transit time of the electron. This scenario is based on the assumption that there are no mobile or fixed charges around the electron as it approaches the metal-semiconductor interface. The electron and the induced surface charges are shown in Fig. 3.8 :

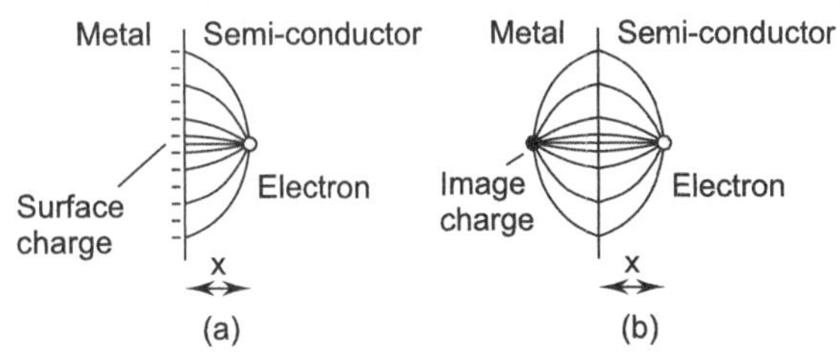

Fig. 3.7 *(a) Field lines and surface charges due to an electron in close proximity to a perfect conductor and (b) the field lines and image charge of an electron*

It can be shown that the electric field in the semiconductor is identical to that of the carrier itself and another carrier with opposite charge at equal distance but on the opposite side of the interface. This charge is called the image charge. The difference between the actual surface charges and the image charge is that the fields in the metal are distinctly different. The image charge concepts is justified on the basis that the electric field lines are perpendicular to the surface a perfect conductor, so that, in the case of a flat interface, the mirror image of the field lines provides continuous field lines across the interface.

The barrier lowering depends on the square root of the electric field at the interface and is calculated from :

$$\Delta\phi_B = \sqrt{\frac{q\varepsilon_{max}}{4\pi\varepsilon_s}} \quad \text{... (3.16)}$$

3.6 SCHOTTKY DIODE CURRENT

The current across a metal-semiconductor junction is mainly due to majority carriers. Three distinctly different mechanisms exist— diffusion of carriers from the semiconductor into the metal, thermionic emission of carriers across the Schottky barrier and quantum-mechanical tunneling through the barrier. The diffusion theory assumes that the driving force is distributed over the length of the depletion layer. The thermionic emission theory on the other hand postulates that only energetic carriers, those, which have an energy equal to or larger than the conduction band energy at the metal-semiconductor interface, contribute to the current flow. Quantum-mechanical tunneling through the barrier takes into account the wave-nature of the electrons, allowing them to penetrate through thin barriers to a given junction, a combination of all three mechanisms could exist. However, typically one finds that only one limits the current, making it the dominant current mechanism.

The analysis reveals that the diffusion and thermionic emission currents can be written in the following form :

$$J_n = qvN_c \exp\left(-\frac{\phi_B}{V_t}\right)\left(\exp\left(\frac{V_a}{V_t}\right)-1\right) \quad \text{... (3.17)}$$

This expression states that the current is the product of the electronic charge, q, a velocity, v, and the density of available carriers in the semiconductor located next to the interface. The velocity equals the mobility multiplied with the field at the interface for the diffusion current and the Richardson velocity for the thermionic emission current. The minus one term ensures that the current is zero if no voltage is applied as in thermal equilibrium any motion of carriers is balanced by a motion of carriers in the opposite direction.

The tunneling current is of a similar form, namely :

$$J_n = qv_R\, n\, \Theta \quad \text{... (3.18)}$$

where v_R is the Richardson velocity and n is the density of carriers in the semiconductor. The tunneling probability term, Θ, is added since the total current depends on the carrier flux arriving at the tunnel barrier multiplied with the probability, Θ, that they tunnel through the barrier.

3.6.1 Diffusion Current

This analysis assumes that the depletion layer is large compared to the mean free path, so that the concepts of drift and diffusion are valid. The resulting current density equals:

$$J_n = \frac{q^2 D_n N_c}{V_t} \sqrt{\frac{2q(\phi_i - V_a) N_d}{\varepsilon_s}} \exp\left(-\frac{\phi_B}{V_t}\right)\left[\exp\left(\frac{V_a}{V_t}\right) - 1\right] \quad \text{... (3.19)}$$

The current therefore depends exponentially on the applied voltage, V_a, and the barrier height, ϕ_B. The prefactor can more easily be understood if one rewrites it as a function of the electric field at the metal-semiconductor interface, ε_{max} :

$$\varepsilon_{max} = \sqrt{\frac{2q(\phi_i - V_a) N_d}{\varepsilon_s}} \quad \text{... (3.20)}$$

yielding :

$$J_n = q \cdot \mu_n \varepsilon_{max} N_c \exp\left(-\frac{\phi_B}{V_t}\right)\left[\exp\left(\frac{V_a}{V_t}\right) - 1\right] \quad \text{... (3.21)}$$

3.6.2 Thermionic Emission

The thermionic emission theory assumes that electrons, which have an energy larger than the top of the barrier, will cross the barrier provided they move towards the barrier. The actual shape of the barrier is hereby ignored. The current can be expressed as :

$$J_{MS} = A^* T^2 e^{-\phi_B/V_t}\left(e^{V_a/V_t} - 1\right) \quad \text{... (3.22)}$$

where $A^* = \frac{4\pi q m^* k^2}{h^3}$ is the Richardson constant and ϕ_B is the Schottky barrier height.

The expression for the current due to thermionic emission can also be written as a function of the average velocity with which the electrons at the interface approach the barrier. This velocity is referred to as the Richardson velocity given by :

$$v_R = \sqrt{\frac{kT}{2\pi m}} \quad \text{... (3.23)}$$

So that the current density becomes :

$$J_n = q v_R N_c \exp\left(-\frac{\phi_B}{V_t}\right)\left[\exp\left(\frac{V_a}{V_t}\right) - 1\right] \quad \text{... (3.24)}$$

3.6.3 Tunneling

The tunneling current is obtained from the product of the carrier charge, velocity and density. The velocity equals the Richardson velocity, the velocity with which on average the carriers approach the barrier. The carrier density equals the density of available electrons, n, multiplied with the tunneling probability, Θ, yielding:

$$J_n = qv_R \, n \, \Theta \qquad \text{... (3.25)}$$

Where the tunneling probability is obtained from :

$$\Theta = \exp\left(-\frac{4}{3}\frac{\sqrt{2qm^*}}{\hbar}\frac{\phi_B^{3/2}}{\varepsilon}\right) \qquad \text{... (3.26)}$$

and the electric field equals $\varepsilon = \dfrac{\phi_B}{L}$.

The tunneling current therefore depends exponentially on the barrier height, ϕ_B to the $\frac{3}{2}$ power.

3.7 SCHOTTKY DIODES

3.7.1 Metal-Semiconductor Contact at Zero Bias

Electrons in the conduction band of a crystal can be viewed as sitting in a potential box formed by the crystal boundaries (see Fig. 3.8). This potential box for electrons is usually deeper in a metal than in a semiconductor. If a metal and a semiconductor are brought together into a close proximity, some electrons from the metal will move into the semiconductor and some electrons from the semiconductor will move into the metal. However, since the barrier for the electron escape from the metal is higher, more electrons will transfer from he semiconductor into the metal than in the opposite direction. At thermal equilibrium, the metal will be charged negatively, and the semiconductor will be charged positively, forming a dipole layer that is very similar to that in a p^+-n junction. The Fermi level will be constant throughout the entire metal-semiconductor system, and the energy band diagram in the semiconductor will be similar to that for an n-type semiconductor in a p^+-n junction (see Fig. 3.9).

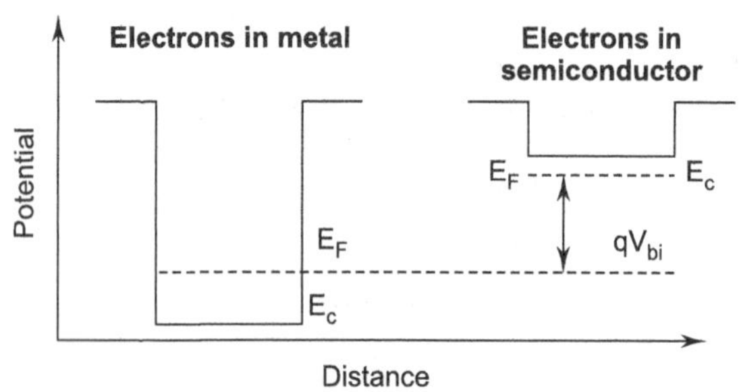

Fig. 3.8 *Schematic energy diagram for electrons in conduction bands of a metal and of a semiconductor.*

Energies ϕ_m and ϕ_s shown in Fig. 3.9 are called the metal and the semiconductor **work functions.** The **work function** is equal to the difference between the **vacuum level** (which is defined as a free electron energy in vacuum) and the Fermi level. The **electron affinity** of the semiconductor, X_s (also shown in Fig. 3.9), corresponds to the energy separation between the vacuum level and the conduction band edge of the semiconductor.

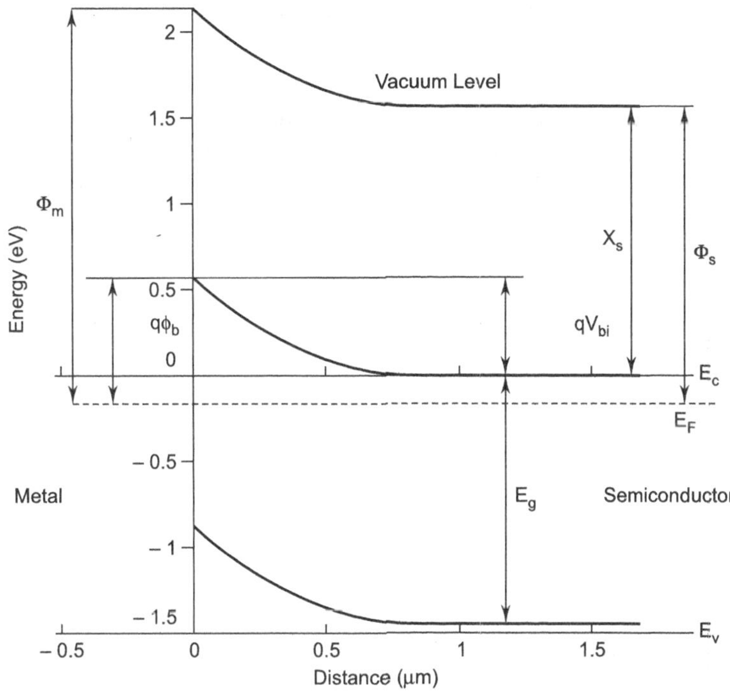

Fig. 3.9 *Simplified energy diagram of GaAs meal-semiconductor barrier $q\ \phi_b$ is the barrier height (0.75 eV), X_s is the electron affinity in the semiconductor, ϕ_s and ϕ_m are the semiconductor and the metal work functions, and V_{bi} (0.591 V) is the built-in voltage. Donor concentration in GaAs is 10^{15} cm^{-3}*

A metal-semiconductor diode is called a **Schottky diode**. In the idealized picture of the Schottky junction shown in Fig. 3.10, the energy barrier between the semiconductor and the metal is

$$q\ \phi_b = \Phi_m - X_s \qquad \text{... (3.27)}$$

Since $\Phi_m > \Phi_s$ the metal is charged negatively. The positive net space charge in the semiconductor leads to a band bending

$$q\ V_{bi} = \Phi_m - \Phi_s \qquad \text{... (3.28)}$$

where V_{bi} is called the **built-in voltage,** in analogy with the corresponding quantity in a *p-n* junction. Note that $q\ V_{bi}$ is also identical to the difference between the Fermi levels in the metal and the semiconductor when separated by a large distance (no exchange of charge) ; see Fig. 3.9.

However, equation (3.27) and Fig. 3.9 are not quite correct. In reality, a change in the metal work function, Φ_m, is not equal to the corresponding change in the barrier height, ϕ_b, as predicted by equation (3.27). In actual Schottky diodes, ϕ_b increases with an increase in Φ_m but only by 0.1 to 0.3 eV when Φ_m increases by 1 to 2 eV. Even though a detailed and accurate understanding of Schottky barrier formation remains a challenge, many properties of Schottky barriers may be understood independently of the exact mechanism determining the barrier height. In other words, we can simply determine the effective barrier height from experimental data. Usually, as a crude and empirical rule of thumb, we can assume that the Schottky barrier height for an n-type semiconductor is close to $\frac{1}{2}$ and $\frac{2}{3}$ of the energy gap.

In a Schottky diode, the semiconductor band diagram looks very similar to that of an n-type semiconductor in a p^+-n diode. Hence, the variation of the space charge density, ρ, the electric field, F, and he potential, φ, in the semiconductor near the metal-semiconductor interface can be found using the depletion approximation :

$$R_c = \frac{\rho_c}{S} \qquad \text{... (3.29)}$$

$$R_c = \frac{R_{cm}}{W} \qquad \text{... (3.30)}$$

$$R_{n,n+1} = 2R_c + R_{sq}\frac{L_{n,n+1}}{W} \qquad \text{... (3.31)}$$

(Here $x = 0$ corresponds to the metal-semiconductor interface.) The depletion layer width, x_n, at zero bias is given by

$$R_{sq} = \frac{1}{\sigma t} \qquad \text{... (3.32)}$$

3.7.2 Schottky Diode Under Bias

Forward bias corresponds to a positive voltage applied to the metal with respect to the semiconductor. Just as for a p^+-n junction, the depletion width under small forward bias and reverse bias may be obtained by substituting V_{bi} with $V_{bi} - V$, where V is the applied voltage. As illustrated in Fig. 3.10, the application of a forward bias decreases the potential barrier for electrons moving from the semiconductor into the metal. Hence, the current-voltage characteristic of a Schottky diode can be described by a **diode equation**, similar to that for a p-n junction diode :

$$I = I_s\left[\exp\left(\frac{V - IR_s}{\eta V_{th}}\right) - 1\right] \qquad \text{... (3.33)}$$

where I_s is the saturation curent, R_s is the series resistance, $V_{th} = \frac{k_B T}{q}$ is the thermal voltage, and η is the ideality factor (η typically varies from 1.02 to 1.6).

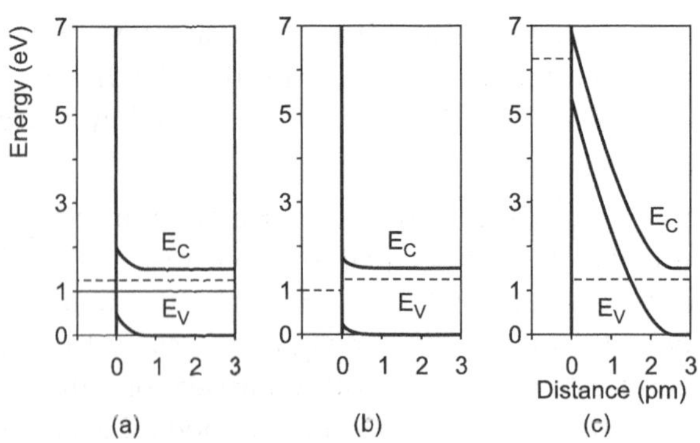

Fig. 3.10 *Band diagram for a GaAs Schottky barrier diode at (a) zero bias, (b) 0.2 V forward bias, and (c) 5 V reverse bias. Dashed line shows the position of the Fermi level in the metal (x < 0) and in the semiconductor (x > 0)*

3.7.3 Thermionic Emission

The diode saturation current, I_s, is typically much larger for Schottky barrier diodes than in *p-n* junction diodes since the Schottky barrier height is smaller than the barrier height in *p-n* junction diodes. In a p-n junction, the height of the barrier separating electrons in the conduction band of the *n*-type region from the bottom of the conduction band in the *p*-region is on the order of the energy gap. A typical Schottky barrier height is only about two thirds of the energy gap or less, as mentioned above. Also, the mechanism of the electron conduction is different. One can show that the saturation current density in a Schottky diode with a relatively low doped semiconductor is given by :

$$J_{ss} = A^* T^2 \exp\left(-\frac{\phi_B}{k_B T}\right) \qquad \text{... (3.34)}$$

where A^* is called the Richardson constant. For a conduction band minimum with spherical surfaces of equal energy (such as the G minimum in GaAs),

$$A^* = \alpha \frac{m_n q k_B^2}{2\pi^2 \hbar^3} \approx 120\, \alpha \frac{m_n}{m_e}\left(\frac{A}{cm^2\, K^2}\right) \qquad \text{... (3.35)}$$

where m_n is the effective mass and α is an empirical factor on the order of unity. The Schottky diode model described by equations (3.34) and (3.35) is called the thermionic emission model. For Schottky barrier diodes of Si, $A^* = 96\ A/(cm^2\, K^2)$. For GaAs, $A^* = 4.4\ A/(cm^2\, K^2)$.

The basic assumption of the thermionic model is that electrons have to pass over the barrier in order to cross the boundary between the metal and the semiconductor. Hence, to find the saturation current, we have to estimate the number of electrons passing over the barrier and their velocities. The number of electrons, N (E) *d*E, having energies between E and E + *d*E is proportional to the product of the Fermi-Dirac distribution function, *f* (E), and the number of states in this energy interval, *g* (E) *d*E, where *g* (E) is the density of states :

$$N(E)\, dE = g(E) f(E)\, dE \qquad \text{... (3.36)}$$

[N (E) = *dn* (E) *d*E where *n* (E) is the number of electrons in the conduction band with energies higher than E. At high energies, the Fermi-dirac occupation function is very close to the Boltzmann distribution function :

$$f_n(E) = \frac{1}{1 + \exp\left(\frac{E - E_F}{k_B T}\right)} \approx \exp\left(\frac{E_F - E}{k_B T}\right) \qquad \text{... (3.37)}$$

The next step should be to multiply the number of the electrons, N (E) *d*E, in the energy interval from E to E + *d*E by the velocity of these electrons. We have to account for different directions of the electron velocities and integrate over energies higher than the barrier height in order to determine the flux of the electrons coming from the semiconductor into the metal. Finally, we deduct the flux of the electrons coming from the metal into the semiconductor. The difference between these two fluxes will be proportional to the current density predicted by the thermionic model. However, we can take a much simpler route if we are interested in understanding the physics of the thermionic model. To this end, let us consider a Schottky diode under a strong reverse bias when V is negative and $-V >> \eta\, k_B T$. Then $I = -I_s$, and the band diagram looks like that shown in Fig. 3*c*. In this case, the energy difference betwen the Fermi level in the semiconductor and the top of the barrier is so large that practically no electrons are available to come from the semiconductor into the metal. However, the Fermi level in the metal is much closer to the top of the barrier, and electrons still come from the metal into the semiconductor. The flux of these electrons constitutes the saturation current. In order to estimate this flux, we should recall that the density of states is a relatively slow function of energy [g (E) is proportional to $(E - E_c)^{1/2}$; compared to the distribution function, which decreases by exp (1) ≈ 2.718 each time E increases by k_B T. Hence, the largest contribution into the electron flux will come from the electrons that are a few k_B T above the barrier. The number of such electrons will be proportional to the effective density of states in the semiconductor.

$$N_c = 2\left(\frac{m_n k_B T}{2\pi\hbar^2}\right)^{3/2} \qquad \text{... (3.38)}$$

and to $\exp\left(-\frac{\phi_b}{k_B T}\right)$. Their velocity in the direction perpendicular to the metal semiconductor interface is proportional to the thermal velocity

$$v_{thnx} = \sqrt{\frac{k_B T}{m_n}} \qquad \text{... (3.39)}$$

Hence, the saturation current density is given by

$$j_{ss} = Cq\,N_c \exp\left(-\frac{\phi_b}{k_B T}\right) v_{thnx} = 2C\left(\frac{m_n k_B T}{2\pi\hbar^2}\right)^{3/2} \sqrt{\frac{k_B T}{m_n}} \exp\left(-\frac{\phi_b}{k_B T}\right)$$

$$= \sqrt{\frac{1}{2\pi^3}}\, C\, \frac{m_n k_B^2 T^2}{\hbar^3} \exp\left(-\frac{\phi_b}{k_B T}\right) \qquad \text{... (3.40)}$$

where C is a numerical constant of the order of unity. With a proper choice of C, this equation coincides with equations (3.34) and (3.35).

3.7.4 Thermionic-field Emission

In relatively high doped semiconductors, the depletion region becomes so narrow that electrons can tunnel through the barrier near the top (see Fig. 3.4*b*). This process is called **thermionic-field emission**. In order to understand thermionic-field emission, we have to recall once again that the number of electrons with energies above a given energy E decreases exponentially with energy as $\exp\left[-\frac{E}{(k_B T)}\right]$. On the other hand, the barrier transparency increases exponentially with the decrease in the barrier width. Hence, as the doping increases and the barrier becomes thinner, the dominant electron tunneling path occurs at lower energies than the top of the barrier (see Fig. 3.11*b*).

In degenerate semiconductors, especially in semiconductors with a small electron effective mass such as GaAs, electrons can tunnel through the barrier near or at the Fermi level, and the tunneling current is dominant. This mechanism is called **field emission** (see Fig. 3.11*c*).

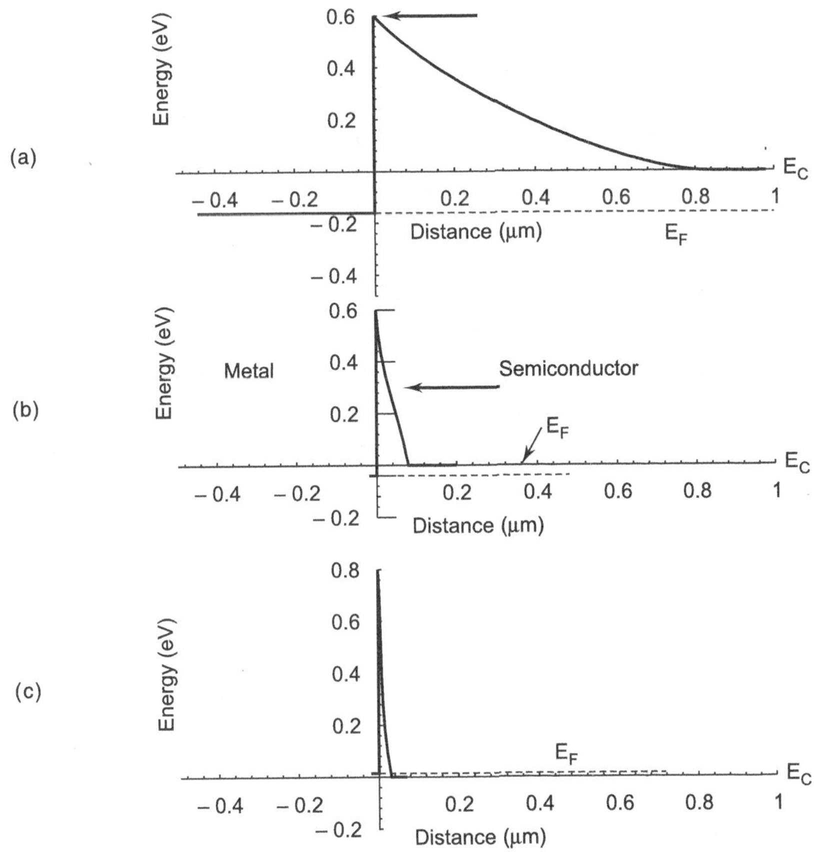

Fig. 3.11

Band diagrams of Schottky barrier junctions for GaAs for doping levels $N_d = 10^{15}$ cm^{-3} (top graph), $N_d = 10^{17}$ cm^{-3} (middle graph), and $N_d = 10^{18}$ cm^{-3} (bottom graph). Arrows indicate electron transfer across the barrier under forward bias. At very low doping levels, electrons go over the barreir closer to the top of the barrier (this process is called thermionic emission). At moderated doping levels, electrons tunnel across the barrier closer to the top of the barrier (this process is called thermionic-field emission). In highly doped degenerate semiconductors, electrons near the Fermi level tunnel across a very thin depletion region (this process is called field emission).

The current-voltage characteristic of a Schottky diode in the case of thermionic-field emission can be calculated using the same approach as for the thermionic model, except that in thermionic-field emission case, we have to evaluate the product of the tunneling transmission coefficient and the number of electrons at a given energy as a function of energy and integrate over the states in the conduction band. Such a calculation [see Rhoderick and Williams (1988)] yields the following expression for the current density in the thermionic-field emission regime under forward bias :

$$j = j_{tf} \exp\left(\frac{qV}{E_o}\right) \quad \text{... (3.41)}$$

where $$E_o = E_{oo} \coth\left(\frac{E_{oo}}{k_B T}\right) \quad \text{... (3.42)}$$

$$E_{oo} = \frac{qh}{4\pi}\sqrt{\frac{N_d}{m_n \varepsilon_s}} = 1.85 \times 10^{-11}\left[\frac{N_d(\text{cm}^{-3})}{(m_n/m_o)(\varepsilon_s/\varepsilon_o)}\right](\text{eV}) \quad \text{... (3.43)}$$

$$j_{stf} = \frac{A^* T\sqrt{\pi E_{oo}\left(\phi_b - qV - E_c + E_{Fn}\right)}}{k_B \cosh\left(E_{oo}/k_B T\right)} \exp\left[-\frac{E_c - E_{Fn}}{k_B T} - \frac{\left(\phi_b - E_c + E_{Fn}\right)}{E_o}\right] \quad \text{... (3.44)}$$

In GaAs Schottky diodes, the thermionic-field emission becomes important for $N_d > 10^{17}$ cm^{-3} at 300 K and for $N_d > 10^{16}$ cm^{-3} at 77 K. In silicon, the corresponding values of N_d are several times larger. The forward *j*-V characteristics are shown in Fig. 3.12.

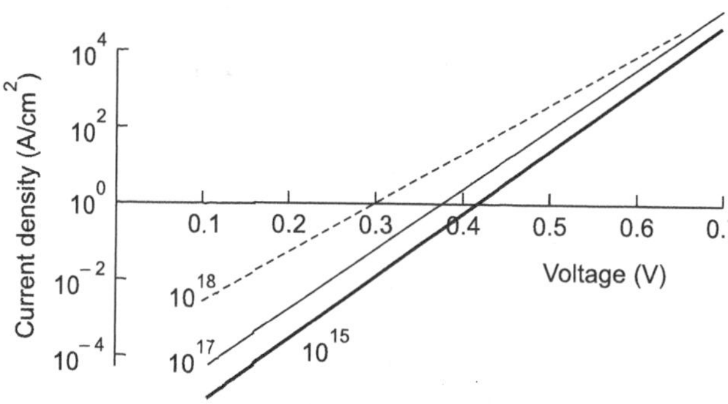

Fig. 3.12 *Forward j-V characteristics of GaAs Schottky diodes doped at 10^{15}, 10^{17}, and 10^{18} cm^{-3} (curves are marked accordingly) at T = 300 K*

The resistance of the Schottky barrier in the field emission regime is quite low. Therefore metal-n^+ contacts are used as ohmic contacts. The specific contact resistance, ρ_c, decreases with the increase in the doping level of the semiconductor. (This resistance may vary from 10^{-3} Ω cm^2 to 10^{-7} Ω cm^2 or even smaller depending on semiconductor material, doping level, contact metal, and ohmic contact fabrication technology).

A Schottky diode is a majority carrier device, where electron-hole recombination is usually not important. Hence, Schottky diodes have a much faster response under forward bias conditions than p-n junction diodes. Therefore, Schottky diodes are used in applications where the speed of a response is important, for example, in microwave **detectors, mixers,** and **varactors.**

SCHOTTKY DIODE

- Structures of Schottky diode
  - Schottky contact/semiconductor/ohmic contact
  - What Schottky and ohmic metal systems do you apply for p-type Si ?

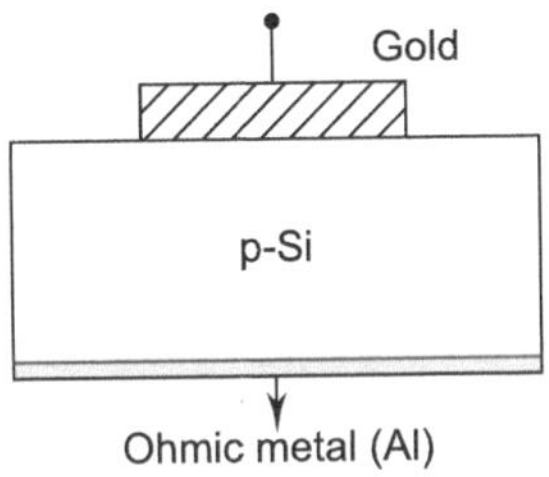

Fig. (i)

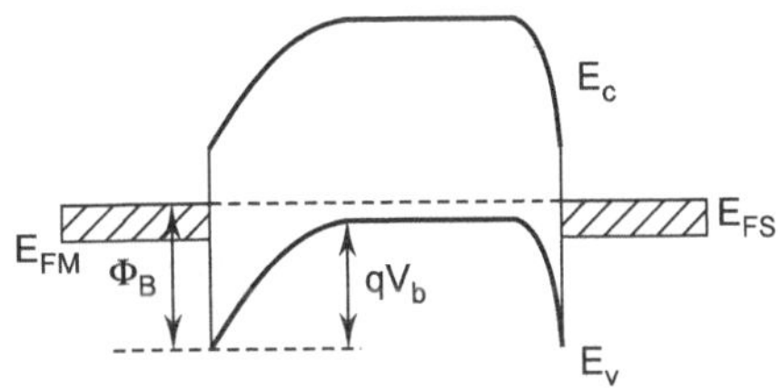

Fig. (ii)

SCHOTTKY DIODE ELECTROSTATICS

- Built-in Voltage

$$V_b = [\phi_B - (E_c - E_{FS})_{FB}] / q$$

- Charge distribution

$$p = \begin{cases} qN_D & ; \quad 0 \le x \le W \\ 0 & ; \quad x \le W \end{cases}$$

- Electric field

$$E(x) = -\frac{qN_D}{\varepsilon_s}(W - x)$$

- Potential

$$V(x) = -\frac{qN_D}{2\varepsilon_s}(W - x)^2$$

- Depletion width

$$W = \left[\frac{2\varepsilon_s}{qN_D}(V_b - V_a)\right]^{1/2}$$

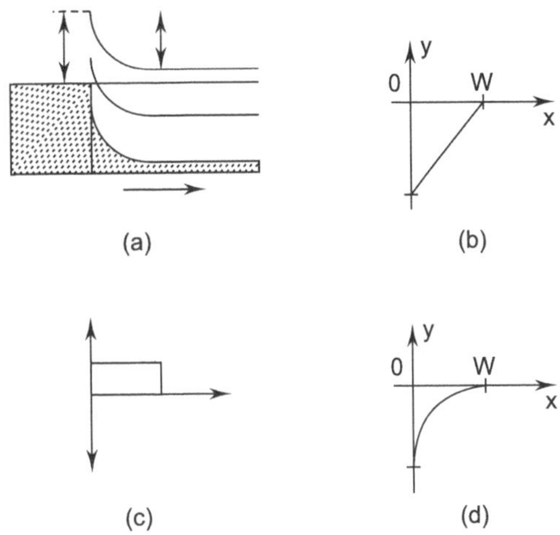

Fig. (iii)

I-V CHARACTERISTICS

- Relatively low barrier for electrons in n-side
- Dominant current from the electron injection from n-side : Themionic emission current.
- At further forward bias, recombination and injection become important.
- At reverse bias, electron injection from metal is dominant and carner generation in the semiconductor depletion region is negligible

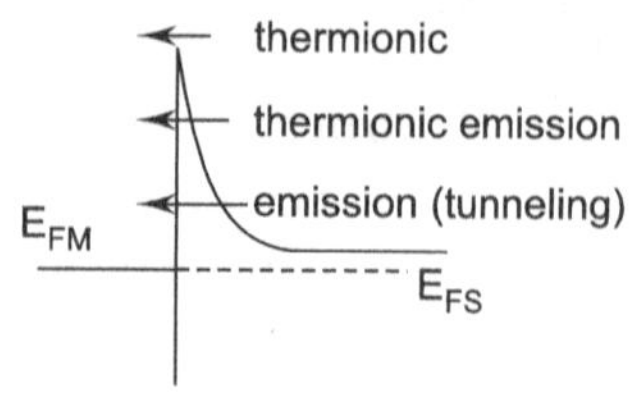

Fig. (iv)

- Curent density from S- to M- is known as the form of

$$J_{s\to w} = A^* T^2 \exp\left(-\frac{\phi_E}{kT}\right)\exp\left(\frac{qV_a}{kT}\right)$$

where A^* = Richardson's constant

- Since $I_{S\to A} = I_{M\to S}$ at equilibrium ($V_a = 0$)

$$J_{M\to S} = A^* T^2 \exp\left(-\frac{\phi_E}{kT}\right)$$

- Therefore,

$$s = j_s\left[\exp\left(-\frac{qV_a}{kT}\right) - 1\right]$$

with $J_s = A^* T^2 \exp\left(-\frac{\phi_B}{kT}\right)$

$$A^* = 4pqm^* k^2 / h^3$$

$\therefore$ $m^*/m_0 = 0.66$ for 2.2 for p- and n-type Si.

- In general

$$J = J_s\left[\exp\left(-\frac{qV_a}{nkT}\right) - 1\right]$$

with n = ideality factor

I-V CHARACTERISTICS

- Important I-V characteristics are :
- Turn-on voltage
- Ideality factor
- Breakdown voltage
- Series resistance

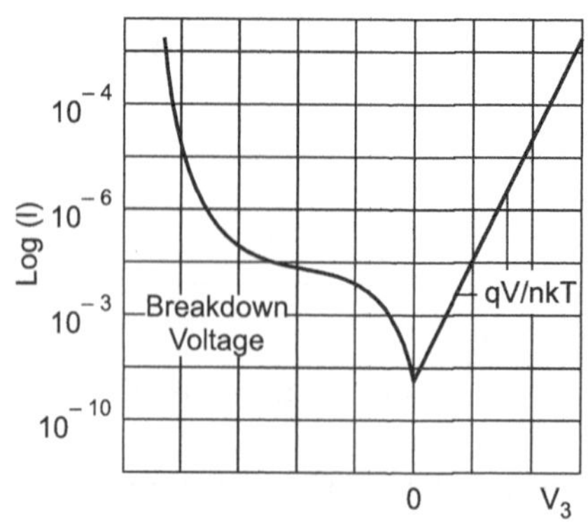

Fig. (v)

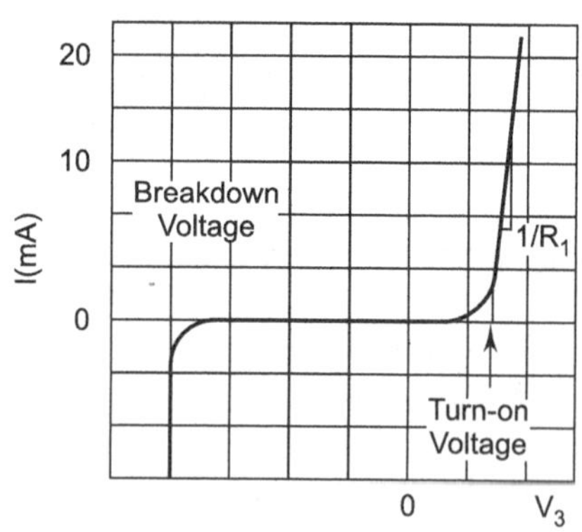

Fig. (vi)

C-V CHARACTERISTICS

- Depletion region only in semiconductor
- Capacitance

$$C = \frac{\varepsilon_s A}{W}$$

$$= \varepsilon_s AI\left[\frac{2\varepsilon_s}{qN_D}(V_b - V_a)\right]^{1/2}$$

- For uniform doped semiconductor

$$\frac{1}{C^2} = \frac{2}{qN_D . \varepsilon_s A^2}(V_b - V_a)$$

- This plot give out about the doping level and the built-in potential.
- Combined with calculated Fermi energy, barrier height can be obtained.

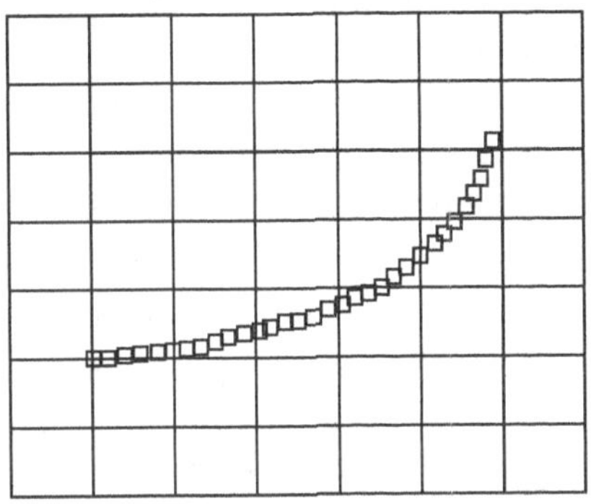

Fig. (vii)

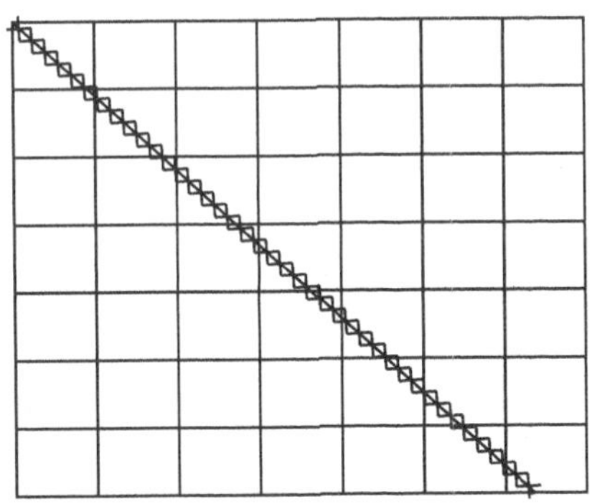

Fig. (viii)

3.8 OHMIC CONTACTS

In the case of a *p-n* diode, for example, contacts have to be provided to both *p*-type and *n*-type regions of the device in order to connect the diode to an external circuit. These contacts have to be as unobtrusive as possible, so that the current flowing through a semiconductor device and, hence, through the contacts, leads to the smallest parasitic voltage drop possible. Whatever voltage drop does occur across the contact has to be proportional to the current so that the contacts do not introduce uncontrollable and unexpected nonlinear elements into the circuit. Since such contacts satisfy Ohm's law, they are usually called ohmic contacts.

As was discussed, a contact between a metal and a semiconductor is typically a Schottky barrier contact. However, if the semiconductor is very highly doped, the Schottky barrier depletion region becomes very thin, as illustrated in Fig. 4. At very high doping levels, a thin depletion layer becomes quite transparent for electron tunneling. This suggests that a pratical way to make a good ohmic contact is to make a very highly doped semiconductor region between the contact metal and the semiconductor.

It may have been better to use a metal with a work function, Φ_m, which is equal to or smaller than the work function of a semiconductor, Φ_s. However, for most semiconductors, it is difficult to find such a metal acceptable for practical contacts.

Current-voltage characteristics of a Schottky barrier diode and of an ohmic contact are compared in Fig. (i). As was mentioned above, a good ohmic contact should have a linear current-voltage characteristic and a very small resistance that is negligible compared to the resistance of the active region of the semiconductor device. An ohmic contact with the I-V characteristic shown in Fig. (ii) does not satisfy fully these conditions since the voltage drop across this contact is not negligibly small compared with the voltage drop across the Schottky diode at moderate current densities above 0.1 kA/cm^2.

As was discussed, the barrier between a metal and a semiconductor is usually smaller for semiconductors with smaller energy gaps. Hence, another way to decrease the contact resistance is to place a layer of a narrow gap highly doped semiconductor material between the active region of the device and the contact metal. Some of the best ohmic contacts to date have been made this way.

A quantitative measure of the contact quality is the **specific contact resistance,** ρ_c, which is the contact resistance of a unit area contact. Depending on the semiconductor material and on the contact quality, ρ_c can vary anywhere from 10^{-3} Ω cm^2 to 10^{-7} Ω cm^2 or even less.

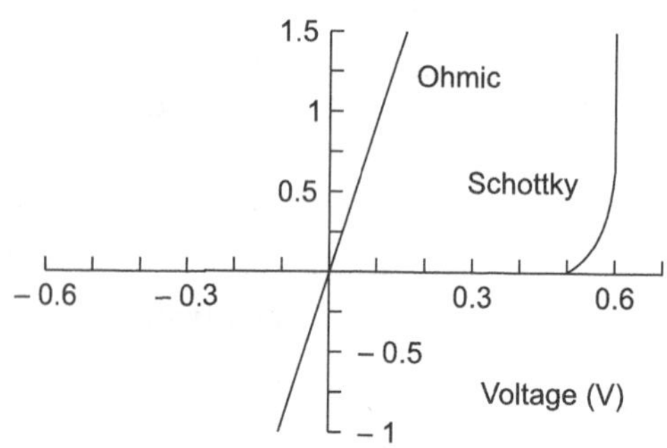

Fig. 3.13 *Current-voltage characteristics of ohmic and Schottky barrier metal-semiconductor contacts to GaAs. (Schottky contact is to GaAs doped at 10^{15} cm^{-3}.) Ohmic contact resistance is 10^4 Ω cm^2*

Most semiconductor devices have either a sandwich structure or a planar structure, as illustrated in Fig. 3.14. The contact resistance of each contact in a sandwich structure contact is given by

$$R_c = \frac{\rho_c}{S} \qquad \text{... (3.45)}$$

A typical current density in a sandwich type device can be as high as 10^4 A/cm^2. Hence, the specific contact resistance of 10^{-5} Ω cm^2 would lead to a voltage drop on the order of 0.1. V. This may be barely acceptable. A larger specific contact resistance of 10^{-4} Ω cm^2 or so would definitely lead to problems, as we can see from Fig. 3.12.

These estimates show that a semiconductor material can become viable for applications in electronic devices only when good ohmic contacts with low contact resistances become available. Often, poor ohmic contacts become a major stumbling block for applications of new semiconductor materials.

3.9 METAL-SEMICONDUCTOR FIELD EFFECT TRANSISTOR (MESFETS)

The Metal-Semiconductor-Field-Effect-Transistor (MESFET) consists of a conducting channel positioned between a source and drain contact region as shown in the Fig. 3.6.1. The carrier flow from source to drain is controlled by a Schottky metal gate. The control of the channel is obtained by varying the depletion layer width underneath the metal contact which modulates the thickness of the conducting channel and therby the current between source and drain.

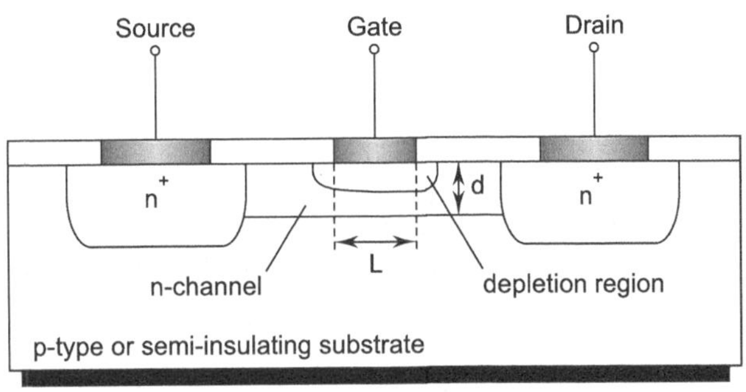

Fig. 3.14 *Structure of a MESFET with gate length, L, and channel thickness, d*

The key advantage of the MESFET is the higher mobility of the carriers in the channel as compared to the MOSFET. Since the carriers located in the inversion layer of a MOSFET have a wavefunction, which extends into the oxide, their mobility—also referred to as surface mobility—is less than half of the mobility of bulk material. As the depletion region separates the carriers from the surface their mobility is close to that of bulk material. The higher mobility leads to a higher current, transconductance nd transit frequency of the device.

The disadvantage of the MESFET structure is the presence of the Schottky metal gate. It limits the forward bias voltage on the gate to the turn-on voltage of the Schottky diode. This turn-on voltage is typically 0.7 V for GaAs Schottky diodes. The threshold voltage therefore must be lower than this turn-on voltage. As a result it is more difficult to fabricate circuits containing a large number of enhancement-mode MESFET.

The higher transit frequency of the MEFSET makes it particularly of interest for microwave circuits. While the advantage of the MESFET provides a superior microwave amplifier or circuit, the limitation by the diode turn-on is easily tolerated. Typically depletion-mode devices are ued since they provide a larger current and larger transconductance and the circuits contain only a few transistors, so that threshold control is not a limiting factor. The buried channel also yields a better noise performance as trapping and release of carriers into and from surface states and defects is eliminated.

The use of GaAs rather than silicon MESFETs provides two more significant advantages—first, the electron mobility at room temperature is more than 5 times larger, while the peak electron velocity is about twice that of silicon. Second, it is possible to fabricate semi-insulating (SI) GaAs substrates, which eliminates the problem of absorbing microwave power in the substrate due to free carrier absorption.

The threshold voltage, V_T, of a MESFET is the voltage required to fully deplete the doped channel layer. This threshold voltage equals :

$$V_T = \phi_i - \frac{qN_d d^2}{2\varepsilon_s} \quad \text{... (3.46)}$$

where ϕ_i is the built-in potential and d is the thickness of the doped region. This threshold voltage can also be written as a function of the pinch-off voltage V_P :

$$V_T = \phi_i - V_P \quad \text{... (3.47)}$$

where the pinch-off voltage equals :

$$V_P = \frac{qN_d d^2}{2\varepsilon_s} \quad \text{... (3.48)}$$

The derivation of the current in a MESFET starts by considering a small section of the device between y and $y + dy$. The current density at that point can be written as a function of the gradient of the channel voltage :

$$J = qnv = qN_d \cdot \mu_n \varepsilon = -qN_d \cdot \mu_n \frac{dV_C(y)}{dy} \quad \text{... (3.49)}$$

The drain current is related to the current density and the part of the MESFET channel that is not depleted.

$$I_D = -JW(d - x_n(y)) \quad \text{... (3.50)}$$

Where the depletion layer width at position y is related to the channel voltage, $V_C(y)$, by :

$$x_n(y) = \sqrt{\frac{2\varepsilon_s(\phi_i - V_G + V_C(y))}{qN_d}} \quad \text{... (3.51)}$$

The equation for the current can now be integrated from source to drain, yielding :

$$I_D = qN_d \mu_n d \frac{W}{L}\left(V_C \Big|_0^{V_D} - \frac{(\phi_i - V_G + V_C)^{3/2}}{\sqrt{V_P}} \Bigg|_0^{V_D} \right) \quad \text{... (3.52)}$$

Since the steady-state current in the device is independent of position, the left hand term equals I_D times L so that :

$$I_D = qN_d \mu_n d \frac{W}{L}\left(V_C \Big|_0^{V_D} - \frac{(\phi_i - V_G + V_C)^{3/2}}{\sqrt{V_P}} \Bigg|_0^{V_D} \right) \quad \text{... (3.53)}$$

Integration results in

$$I_D = q\mu_n N_d d \frac{W}{L}\left[V_D - \frac{2}{3}\left(\frac{(\phi_i - V_G + V_D)^{3/2}}{\sqrt{V_P}} - \frac{(\phi_i - V_G)^{3/2}}{\sqrt{V_P}} \right) \right] \quad \text{... (3.54)}$$

This result is valid as long as the width of the un-depleted channel ($d - x_n$ (y)) is positive, namely for:

$$V_D \le V_G - V_T \quad \text{... (3.55)}$$

This condition also defines the quadratic region of a MESFET. For larger drain voltage, the current saturates and equals that at

$$V_D = V_G - V_T = V_{D,sat} \quad \text{... (3.56)}$$

The corresponding current is the saturation current, $I_{D,\,sat}$:

$$I_{D,sat} = q\mu_n N_d d \frac{W}{L}\left[V_G - V_T - \frac{2}{3}\left(V_P - \frac{(\phi_i - V_G)^{3/2}}{\sqrt{V_P}} \right) \right] \quad \text{... (3.57)}$$

An example of the resulting I-V characteristics is shown in

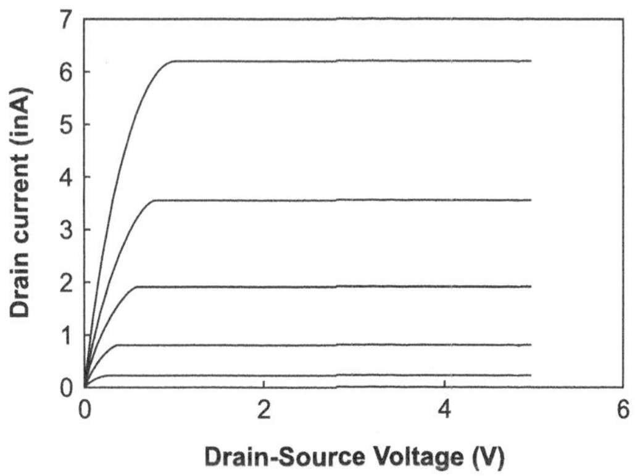

Fig. 3.15 *Drain current versus Drain-Source voltage at a gate-source voltage of 0.2, 0.4, 0.6, 0.8 and 1.0 Volt for a silicon MESFET with built-in potential of 1 V*

The transfer characteristic of a MESFET is shown in Fig. 3.15 and compared to a quadratic expression of the form :

$$I_{D,\,sat} = \mu_n \frac{\varepsilon_s}{\overline{w}} \frac{W}{L} \frac{(V_G - V_T)^2}{2} \qquad \text{... (3.58)}$$

where $\overline{w}$ is the average depletion layer width in the channel layer. The quadratic expression yields the same current at $V_G = \phi_i$ for $\overline{w} = \frac{3d}{8}$. The close fit is at times used to justify using the simpler quadratic equation.

QUESTIONS

- What is a flatband diagram ?
- Define the barrier height of a metal-semiconductor junctiion. Can the barrier height be negative ? Explain.
- Define the built-in potential. Also provide an equation and state the implicit assumption (s).
- How does the energy band diagram of a metal-semiconductor junction change under forward and reverse bias ? How does the depletion layer width change with bias ?
- What is the full depletion approximation ? Why do we need the full depletion approximation?
- What mechanism (s) cause (s) current in a metal-semiconductor junction ?

4

MOS Capacitors

4.1 INTRODUCTION

The primary reason to study the Metal-Oxide-Silicon (MOS) capacitor is to understand the principle of operation as well as the detailed analysis of the Metal-Oxide-Silicon Field Effect Transistor (MOSFET). In this chapter, we introduce the MOS structure and its four different modes of operation, namely accumulation, flatband, depletion and inversion. We then consider the flatband voltage in more detail and present the MOS analysis based on the full depletion approximation. Finally, we analyze and discuss the MOS capacitance.

4.2 STRUCTURE AND PRINCIPLE OF OPERATION

The MOS capacitor consists of a Metal-Oxide-Semiconductor structure as illustrated by Fig. 4.1. Shown in the semiconductor substrate with a thin oxide layer and a top metal contact, referred to as the gate. A second metal layer forms an Ohmic contact to the back of the semiconductor and is called the bulk contact. The structure shown has a p-type substrate. We will refer to this as an n-type MOS or nMOS capacitor since the inversion layer—as discussed in section 4.6.4—contains electrons.

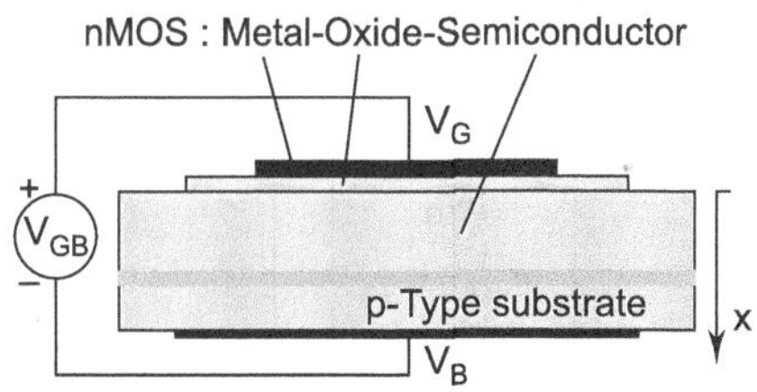

Fig. 4.1 *MOS capacitance structure*

To understand the different bias modes of an MOS capacitor we now consider three different bias voltages. One below the flatband voltage, V_{FB}, a second between the flatband voltage and the threshold voltage, V_T, and finally one larger than the threshold voltage. These bias regimes are called the accumulation, depletion and inversion mode of operation. These three modes as well as the charge distributions associated with each of them are shown in Fig. 4.2.

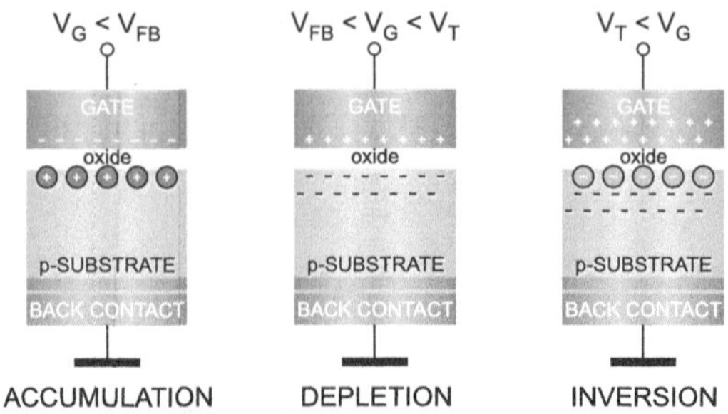

Fig. 4.2 *Charges in an n-type Metal-Oxide-Semiconductor structure (p-type substrate) under accumulation, depletion and inversion conditions*

Accumulation occurs typically for negative voltages where the negative charge on the gate attracts holes from the substrate to the oxide-semiconductor interface. Depletion occurs for positive voltages. The positive charge on the gate pushes the mobile holes into the substrate. Therefore, the semiconductor is depleted of mobile carriers at the interface and a negative charge, due to the ionized acceptor ions, is left in the space charge region. The voltage separating the accumulation and depletion regime is referred to as the *flatband voltage*, V_{FB}. Inversion occurs at voltages beyond the threshold voltage. In inversion, there exists a negatively charged inversion layer at the oxide-semiconductor interface in addition to the depletion-layer. This inversion layer is due to the minority carriers that are attracted to the interface by the positive gate voltage.

The energy band diagram of an *n*-MOS capacitor biased in inversion is shown in Fig. 4.3. The oxide is modeled as a semiconductor with a very large bandgap and blocks any flow of carriers between the semiconductor and the gate metal. The band bending in the semiconductor is consistent with the presence of a depletion layer. At the semiconductor-oxide interface, the Fermi energy is close to the conduction band edge as expected when a high density of electrons is present. The semiconductor remains in thermal equilibrium even when a voltage is applied to the gate. The presence of an electric field does not automatically lead to a non-equilibrium condition, as was also the case for a *p-n* diode with zero bias.

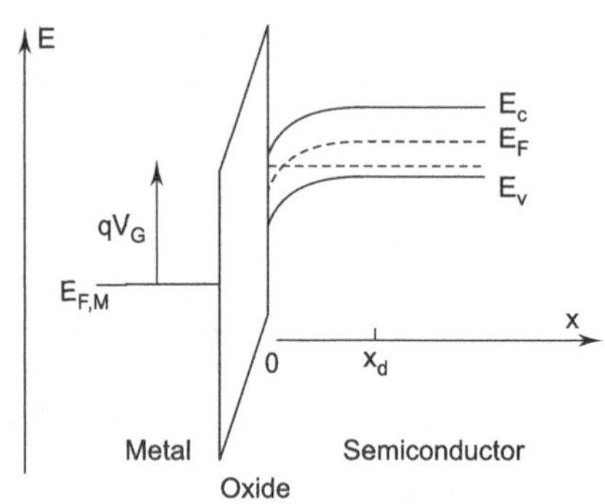

Fig. 4.3 *Energy band diagram of an MOS structure biased in inversion*

In the next sections, we discuss the four modes of operation of an MOS structure : Flatband, Depletion, Inversion and Accumulation. Flatband conditions exist when no charge is present in the semiconductor so that the silicon energy band is flat. Initially we will assume that this occurs at zero gate bias. Later we will consider the actual flatband voltage in more detail. Surface depletion occurs when the holes in the substrate are pushed away by a positive gate voltage. A more positive voltage also attracts electrons (the minority carriers) to the surface, which form the so-called inversion layer. Under negative gate bias, one attracts holes from the p-type substrate to the surface, yielding accumulation.

4.2.1 Flatband Diagram

The flatband diagram is by far the easiest energy band diagram. The term flatband refers to fact that the energy band diagram of the semiconductor is flat, which implies that no charge exists in the semiconductor. The flatband diagram of an aluminium-silicon dioxide-silicon MOS structure is shown in Fig. 4.4. Note that a voltage, V_{FB}, must be applied to obtain this flat band diagram. Indicated on the Fig 4.4. the work function of the aluminium gate is Φ_M, the electron affinity of the oxide is χ_{oxide} and that of silicon is *x* instead of *c*, as well as the bandgap energy of silicon, E_g. The bandgap energy of the oxide is quoted in the literature to be between 8 and 9 electron volt. The reader should also realize that the oxide is an amorphous material and the use of semiconductor parameters for such material can justifiably be questioned.

The flatband voltage is obtained when the applied gate voltage equals the workfunction difference between the gate metal and the semiconductor. If there is a fixed charge in the oxide and/or at the oxide-silicon interface, the expression for the flatband voltage must be modified accordingly.

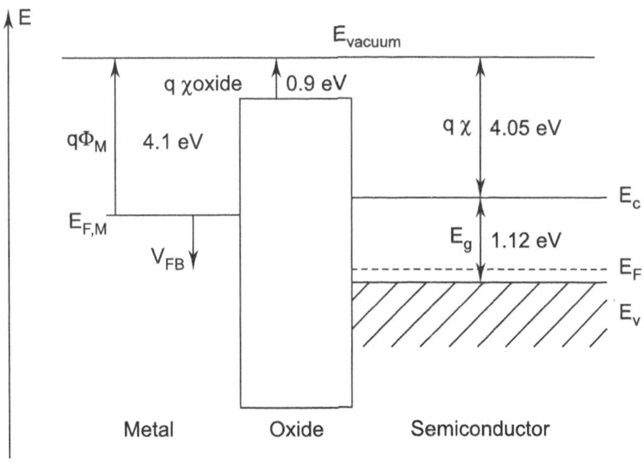

Fig. 4.4 *Flatband energy diagram of a metal-oxide-semiconductor (MOS) structure consisting of an aluminium metal, silicon dioxide and silicon*

4.2.2 Accumulation

Accumulation occurs when one applies a voltage less than the flatband voltage. The negative charge on the gate attracts holes from the substrate to the oxide-semiconductor interface. Only a small amount of band bending is needed to build up the accumulation charge so that almost all of the potential variation is within the oxide.

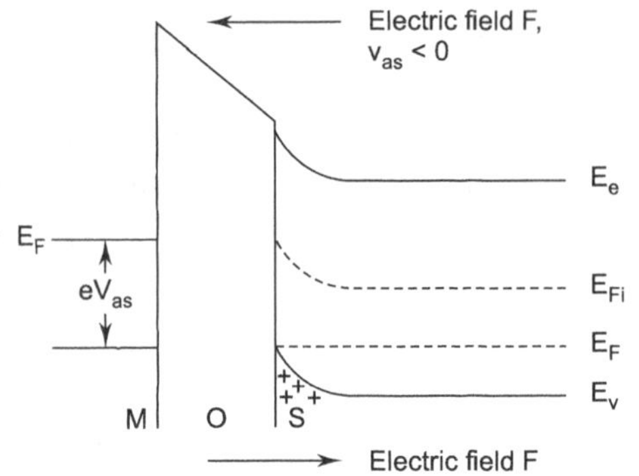

Fig. 4.5 *Negative voltage causes hole accumulation in the p-type semiconductor*

4.2.3 Depletion

As a more positive voltage than the flatband voltage is applied, a negative charge builds up in the semiconductor. Initially this charge is due to the depletion of the semiconductor starting from the oxide-semiconductor interface. The depletion layer width further increases with increasing gate voltage.

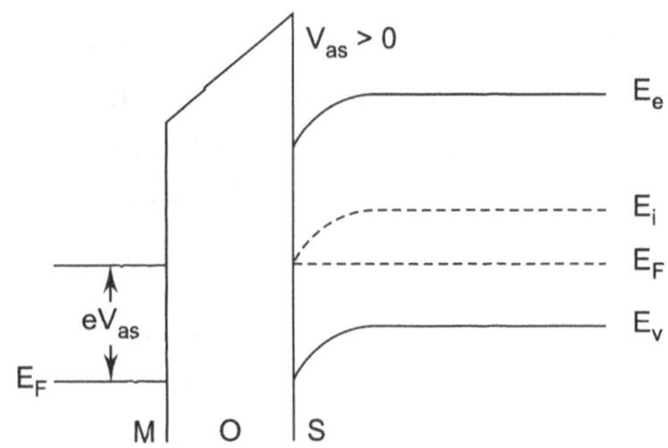

Fig. 4.6 *Positive voltage depletes holes from the semiconductor surface*

4.2.4 Inversion

As the potential across the semi-conductor increases beyond twice the bulk potential, another type of negative charge emerges at the oxide-semiconductor interface : this charge is due to minority carriers, which form a so-called inversion layer. As one further increases the gate voltage, the depletion layer width barely increases further since the charge in the inversion layer increases exponentially with the surface potential.

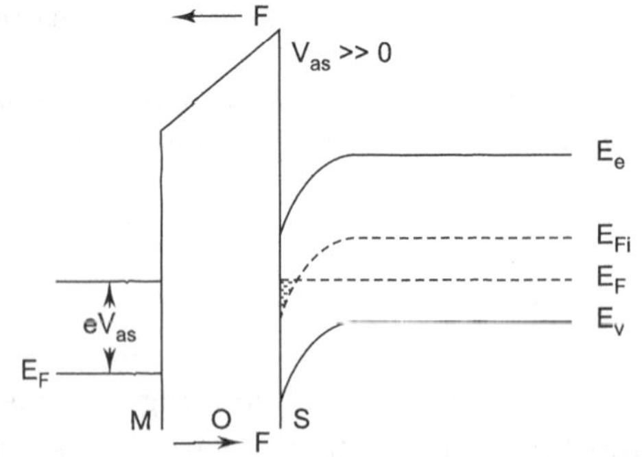

Fig. 4.7 *A larger positive voltage causes inversion – an "n-type" layer at the semiconductor surface*

4.3 MOS ANALYSIS

4.3.1 Flatband Voltage Calculation

If there is no charge present in the oxide or at the oxide-semiconductor interface, the flatband voltage simply equals the difference between the gate metal workfunction, F_M, and the semiconductor workfunction, F_S.

$$V_{FB} = \Phi_M - \Phi_S \quad \text{... (4.1)}$$

The workfunction is the voltage required to extract an electron from the Fermi energy to the vacuum level. This voltage is between three and five Volt for most metals. The actual value of the workfunction of a metal deposited onto silicon dioxide is not exactly the same as that of the metal in vacuum. Figure 4.8 provides experimental values for the workfunction of different metals, as obtained from a measurement on a MOS capacitor, as a function of the measured workfunction in vacuum. The same data is also listed in Table 4.1.

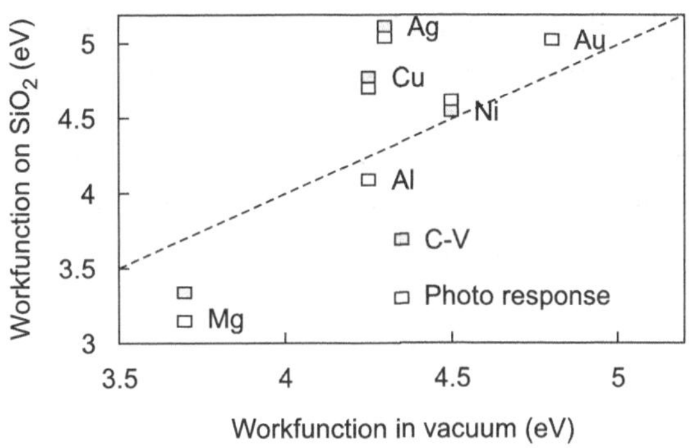

Fig. 4.8 *Workfunction of Magnesium (Mg), Aluminium (Al), Copper (Cu), Silver (Ag), Nickel (Ni) and Gold (Au) obtained from I-V and C-V measurements on MOS structures as a function of the workfunction of those metals measured in vacuum*

Table 4.1 Workfunction of selected metals as measured in vacuum and as obtained from a C-V measurement of an MOS structure

	Ag	*Al*	*Au*	*Cr*	*Mg*	*Ni*
ϕ_M (in vacuum)	4.3	4.25	4.8	4.25	3.7	4.5
ϕ_M (on SiO_2) (C-V)	5.1	4.1	5	4.7	3.35	4.55

The workfunction of a semicondcutor, F_S, requires some more thought since the Fermi energy varies with the doping type as well as with the doping concentration. This workfunction equals the sum of the electron affinity in the semiconductor, *c*, the difference between the conduction band energy and the intrinsic energy divided by the electronic charge in addition to the bulk potential. This is expressed by the following equation :

$$\phi_M - \phi_S = \phi_M - \chi - \frac{E_g}{2q} - V_t \ln\left(\frac{N_a}{n_i}\right) \quad \text{... (4.2)}$$

For MOS structures with a highly doped poly-silicon gate one must also calculate the workfunction of the gate based on the bulk potential of the poly-silicon.

$$\phi_{poly,S} = V_t \ln\left(\frac{N_{a,poly}}{N_a}\right) \qquad p\text{-type polysilicon gate}$$

$$\phi_{poly,S} = V_t \ln\left(\frac{n_i^2}{N_{d,poly} N_a}\right) \qquad n\text{-type polysilicon gate} \qquad \text{... (4.3)}$$

Where $N_{a,poly}$ and $N_{d,poly}$ are the acceptor and donor density of the *p*-type and *n*-type poly-silicon gate respectively.

For a pMOS capacitor, which has an *n*-type substrate with doping density N_d, the workfunction difference equals :

$$\phi_M - \phi_S = \phi_M - \chi - \frac{E_g}{2q} + V_t \ln\left(\frac{N_d}{n_i}\right) \qquad \text{... (4.4)}$$

The flatband voltage of real MOS structures is further affected by the presence of charge in the oxide or at the oxide-semiconductor interface. The flatband voltage still corresponds to the voltage, which, when applied to the gate electrode, yields a flat energy band in the semiconductor. Any charge in the oxide or at the interface affects the flatband voltage. For a charge, Q_i, located at the interface between the oxide and the semiconductor, and a charge density, r_{ox}, distributed within the oxide, the flatband voltage is given by :

$$V_{FB} = \phi_{MS} - \frac{Q_i}{C_{ox}} - \frac{1}{\varepsilon_{ox}} \int_0^{t_{ox}} \rho_{ox}(x)\, x\, dx \qquad \text{... (4.5)}$$

where the second term is the voltage across the oxide due to the charge at the oxide-semiconductor interface and the third term is due to the charge density in the oxide.

The actual calculation of the flatband voltage is further complicated by the fact that charge can move within the oxide. The charge at the oxide-semiconductor interface due to surface states also depends on the position of the Fermi energy.

Since any additional charge affects the flatband voltage and thereby the threshold voltage, great care has to be taken during fabrication to avoid the incorporation of charged ions as well as creation of surface states.

Example. *Calculate the flatband voltage of a silicon nMOS capacitor with a substrate doping $N_a = 10^{17}$ cm^{-3} and an aluminium gate ($F_M = 4.1$ V). Assume there is no fixed charge in the oxide or at the oxide-silicon interface.*

Solution. The flatband voltage equals the work function difference since there is no charge in the oxide or at the oxide-semiconductor interface.

$$V_{FB} = \Phi_{MS} = \Phi_M - \chi - \frac{E_g}{2q} - V_t \ln\frac{N_a}{n_i}$$

$$= 4.1 - 4.05 - 0.56 - 0.026 \times \ln\frac{10^{17}}{10^{10}} = -0.93 \text{ V}$$

The flatband voltages for *n*MOS and *p*MOS capacitors with an aluminium or a polysilicon gate are listed in the table below.

	Aluminium	p^+ poly	n^+ poly
*n*MOS	– 0.93 eV	0.14 eV	– 0.98 eV
*p*MOS	– 0.09 eV	0.098 eV	– 0.14 eV

4.3.2 Inversion Layer Charge

The basis assumption as needed for the derivation of the MOSFET models is that the inversion layer charge is *proportional* with the applied voltage. In addition, the inversion layer charge is zero at and below the threshold voltage as described by :

$$Q_{inv} = C_{ox}(V_G - V_T)$$
$$Q_{inv} = 0 \qquad \text{... (4.6)}$$

The linear proportionality can be explained by the fact that a gate voltage variation causes a charge variation in the inversion layer. The proportionality constant between the charge and the applied voltage is therefore expected to be the gate oxide capacitance. This assumption also implies that the inversion layer charge is located exactly at the oxide-semiconductor interface.

Because of the energy band gap of the semiconductor separating the electrons from the holes, the electrons can only exist if the *p*-type semiconductor is first depleted. The voltage at which the electron inversion-layer forms is referred to as the threshold voltage.

To justify this assumption we now examine a comparison of a numeric solution with equation (4.6) as shown in Fig. 4.9.

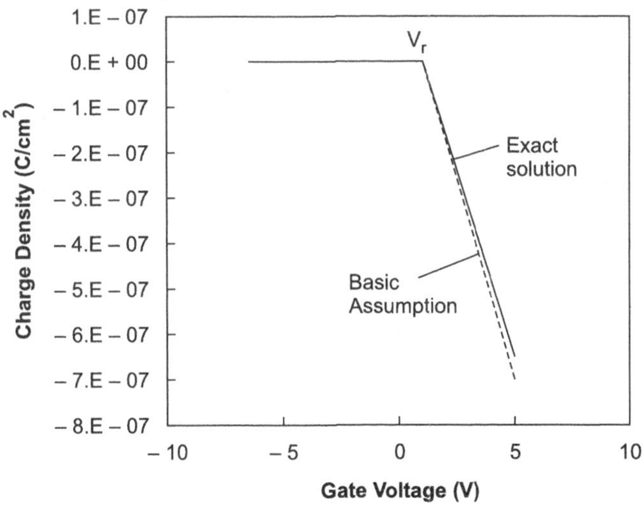

Fig. 4.9 *Charge density due to electrons in the inversion layer of an MOS capacitor. Compared are the analytic solution (solid line) and equation (4.6) (dotted line) for $N_a = 10^{17}\ cm^{-3}$ and $t_{ox} = 20\ nm$.*

While there is a clear difference betwen the curves, the difference is small. We will therefore use our basic assumption when deriving the different MOSFET models since it dramatically simplifies the derivation, be it while losing some accuracy.

4.3.3 Full Depletion Analysis

We now derive the MOS parameters at threshold with the aid of Fig. 4.10. To simplify the anlaysis we make the following assumptions (1) we assume that we can use the full depletion approximation and (2) we assume that the inversion layer charge is zero below the threshold voltage. Beyond the threshold voltage we assume that the inversion layer charge changes linearly with the applied gate voltage.

The derivation starts by examining the charge per unit area in the depletion layer, Q_d. As can be seen in Fig. 4.10 (*a*), this charge is given by :

$$Q_d = -q N_a x_d \qquad \text{... (4.7)}$$

Where x_d is the depletion layer width and N_a is the acceptor density in the substrate. Integration of the charge density then yields the electric field distribution shown in Fig. 4.10 (*b*). The electric field in the semicondcutor at the interface, ε_s, and the field in the oxide equal, s_{ox} :

$$\varepsilon_s = \frac{qN_a x_d}{\varepsilon_s} \text{ and } \varepsilon_{ox} = \frac{qN_a x_d}{\varepsilon_{ox}} \qquad \text{... (4.8)}$$

The electric field changes abruptly at the oxide-semiconductor interface due to the difference in the dielectric constant. At a silicon/SiO_2 interface the field in the oxide is about three times larger since the dielectric constant of the oxide ($e_{ox} = 3.9\ e_0$) is about one third that of silicon ($e_s = 11.9\ e_0$). The electric field in the semiconductor changes linearly due to the constant doping density and is zero at the edge of the depletion region, based on the full depletion approximation.

The potential shown in Fig. 4.10 (*c*) is obtained by integrating the electric field. The potential at the surface, ϕ_s, equals :

$$\phi_s = \frac{qN_a x_d^2}{2\varepsilon_s} \qquad \text{... (4.9)}$$

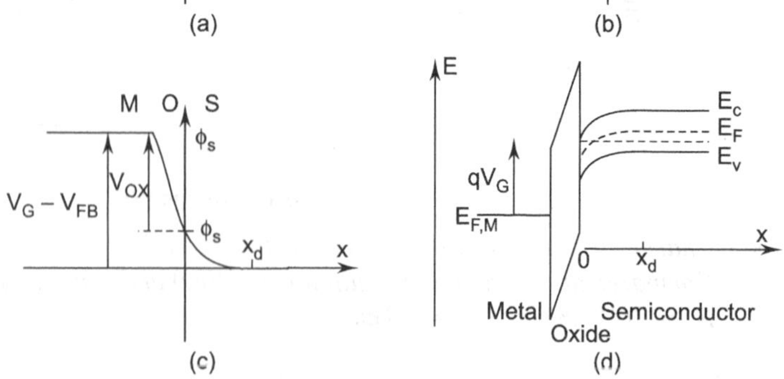

Fig. 4.10 *Electrostatic analysis of an MOS structure. Shown are (a) the charge density, (b) the electric field, (c) the potential and (d) the energy band diagram for an nMOS structure biased in depletion*

The calculated field and potential is only valid in depletion. In accumulation, there is no depletion region and the full depletion approximation does not apply. In inversion, there is an additional charge in the inversion layer, Q_{inv}. This charge increases gradually as the gate voltage is increased. However, this charge is only significant once the electron density at the surface exceeds the hole density in the substrate, N_a. We therefore define the threshold voltage as the gate voltage for which the electron density at the surface equals N_a. This corresponds to the situation where the total potential across the surface equals twice the bulk potential, ϕ_F.

$$\phi_F = V_t \ln \frac{N_a}{n_i} \qquad \text{... (4.10)}$$

The depletion layer in depletion is therefore restricted to this potential range :

$$x_d = \sqrt{\frac{2\varepsilon_s \phi_s}{qN_a}}, \text{ for } 0 \le \phi_s \le 2\phi_F \qquad \text{... (4.11)}$$

For a surface potential larger than twice the bulk potential, the inversion layer charge increases exponentially with the surface potential. Consequently, an increasd gate voltage voltage yields an increased voltage across the oxide while the surface potential remains almost constant. We will therefore assume that the surface potential and the depletion layer width at threshold equal those in inversion. The corresponding expressions for the depletion layer charge at threshold, $Q_{d,\,T}$, and the depletion layer width at threshold, $x_{d,\,T}$, are :

$$Q_{d,T} = -q\,N_a\,x_{d,T} \qquad \text{... (4.12)}$$

$$x_{d,T} = \sqrt{\frac{2\varepsilon_s (2\phi_F)}{qN_a}} \qquad \text{... (4.13)}$$

Beyond threshold, the total charge in the semiconductor has to balance the charge on the gate electrode, Q_M, or :

$$Q_M = -(Q_d + Q_{inv}) \qquad \text{... (4.14)}$$

where we define the charge in the inversion layer as a quantity, which needs to determined but should be consistent with our basic assumption. This leads to the following expression for the gate voltage, V_G :

$$V_G = V_{FB} + \phi_s + \frac{Q_M}{C_{ox}} = V_{FB} + \phi_s - \frac{Q_d + Q_{inv}}{C_{ox}} \qquad \text{... (4.15)}$$

In depletion, the inversion layer charge is zero so that the gate voltage becomes :

$$V_G = V_{FB} + \phi_s + \frac{\sqrt{2\varepsilon_s q N_a \phi_s}}{C_{ox}}, \text{ for } 0 \le \phi_s \le 2\phi_F \qquad \text{... (4.16)}$$

while in inversion this expression becomes :

$$V_G = V_{FB} + 2\phi_F + \frac{\sqrt{4\varepsilon_s q N_a \phi_F}}{C_{ox}} - \frac{Q_{inv}}{C_{ox}} \overset{\Delta}{=} V_T - \frac{Q_{inv}}{C_{ox}} \quad \text{... (4.17)}$$

the third term in (4.3.17) states our basic assumption, namely that any change in gate voltage beyond the threshold requires a change of the inversion layer charge. From the second equality in equation (4.3.17), we then obtain the threshold voltage or :

$$V_T = V_{FB} + 2\phi_F + \frac{\sqrt{4\varepsilon_s q N_a \phi_F}}{C_{ox}} \quad \text{... (4.18)}$$

Example : *Calculate the threshold voltage of a silicon nMOS capacitor with a substrate doping $N_a = 10^{17}$ cm^{-3}, a 20 nm thick oxide ($e_{ox} = 3.9\ e_0$) and an aluminium gate ($F_M = 4.1$ V). Assume there is no fixed charge in the oxide or at the oxide-silicon interface.*

Solution. The threshold voltage equals :

$$V_T = V_{FB} + 2\phi_F + \frac{\sqrt{4\varepsilon_s q N_a \phi_F}}{C_{ox}}$$

$$= -0.93 + 2 \times 0.42$$

$$+ \frac{\sqrt{4 \times 11.9 \times 8.85 \times 10^{-14} \times 1.6 \times 10^{-19} \times 10^{17} \times 0.42}}{3.9 \times 8.85 \times 10^{-14} / 20 \times 10^{-7}}$$

$$= -0.09 \text{ V}$$

Where the flatband voltage was already calculated in example 4.1. The threshold voltage voltages for nMOS and pMOS capacitors with an aluminium or a poly-silicon gate are listed in the table below.

	Aluminium	p^+ poly	n^+ poly
*n*MOS	– 0.09 eV	0.98 eV	– 0.14 eV
*p*MOS	– 0.93 eV	0.14 eV	– 0.98 eV

4.3.4 MOS Capacitance

Capacitance voltage measurements of MOS capacitors provide a wealth of information about the structure, which is of direct interest when one evaluates an MOS process. Since the MOS structure is simple to fabricate, the technique is widely used.

To understand capacitance-voltage measurements one must first be familiar with the frequency dependence of the measurement. This frequency dependence occurs primarily in inversion since a certain time is needed to generate the minority carriers in the inversion layer. Thermal equilibrium is therefore not immediately obtained.

The **low frequency** or quasi-static measurement maintains thermal equilibrium at all times. This capacitance is the ratio of the change in charge to the change in gate voltage, measured while the capacitor is in equilibrium. A typical measurement is performed with an electrometer, which measures the charge added per unit time as one slowly varies the applied gate voltage.

The **high frequency** capacitance is obtained from a small-signal capacitance measurement at high frequency. The bias voltage on the gate is varied slowly to obtain the capacitance versus voltage. Under such conditions, one finds that the charge in the inversion layer does not change from the equilibrium value at the applied d.c. voltage. The high frequency capacitance therefore reflects only the charge variation in the depletion layer and the (rather small) movement of the inversion layer charge.

In this section, we first derive the simple capacitance model, which is based on the full depletion approximation and our basic assumption. The comparison with the exact low frequency capacitance will reveal that the largest error occurs at the flatband voltage. We therefore derive the exact flatband capacitance using the linearized Poisson's equation. Then we discuss deep depletion as well as the non-ideal effects in MOS capacitors.

4.3.4.1 Simple capacitance model

The capacitance of an MOS capacitor is obtained using the same assumptions as those listed in section 4.3.3. The MOS structure is treated as a series connection of two capacitors: the capacitance of the oxide and the capacitance of the depletion layer.

In accumulation, there is no depletion layer. The remaining capacitor is the oxide capacitance, so that the capacitance equals :

$$C_{LF} = C_{HF} = C_{ox}, \text{ for } V_G \le V_{FB} \qquad \text{... (4.19)}$$

In depletion, the MOS capacitance is obtained from the series connection of the oxide capacitance and the capacitance of the depletion layer, or :

$$C_{LF} = C_{HF} = \frac{1}{\dfrac{1}{C_{ox}} + \dfrac{x_d}{\varepsilon_s}}, \text{ for } V_{FB} \le V_G \le V_T \qquad \text{... (4.20)}$$

where x_d is the variable depletion layer width which is calculated from :

$$x_d = \sqrt{\frac{2\varepsilon_s \phi_s}{qN_d}} \qquad \text{... (4.21)}$$

In order to find the capacitance corresponding to a specific value of the gate voltage, we also need to use the relation between the potential across the depletion region and the gate voltage, given by :

$$V_G = V_{FB} + \phi_s + \frac{\sqrt{2\varepsilon_s qN_a \phi_s}}{C_{ox}}, \text{ for } 0 \le \phi_s \le 2\,\phi_F \qquad \text{... (4.16)}$$

In inversion, the capacitance becomes independent of the gate voltage. The low frequency capacitance equals the oxide capacitance since charge is added to and removed from the inversion layer. The high frequency capacitance is obtained from the series connection of the oxide capacitance and the capacitance of the depletion layer having its maximum width, $x_{d,\,T}$. The capacitances are given by:

$$C_{LF} = C_{ox} \text{ and } C_{HF} = \frac{1}{\frac{1}{C_{ox}} + \frac{x_{d,T}}{\varepsilon_s}}, \text{ for } V_G \geq V_T \qquad \text{... (4.22)}$$

The capacitance of an MOS capacitor as calulated using the simple model is shown in Fig. 4.11 The dotted lines represent the simple model while the solid line corresponds to the low frequency capacitance as obtained from the exact analysis.

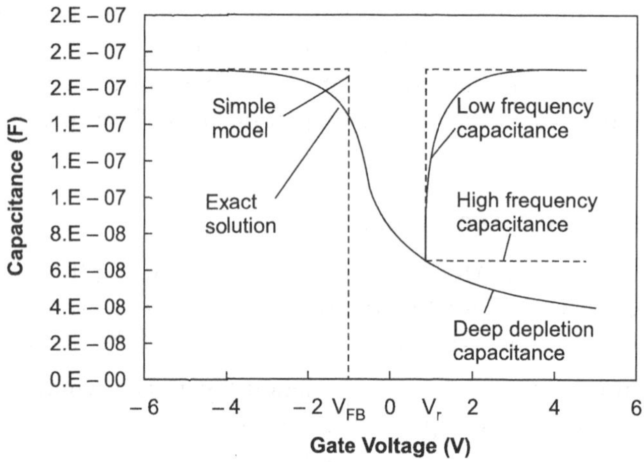

Fig. 4.11 *Low frequency capacitance of an nMOS capacitor. Shown are the exact solution for the low frequency capacitance (solid line) and the low and high frequency capacitance obtained with the simple model (dotted lines)* . $N_a = 10^{17}\ cm^{-3}$ *and* $t_{ox} = 20\ nm$.

4.3.4.2 Calculation of the flat band capacitance

The simple model predicts that the flatband capacitance equals the oxide capacitance. However, the comparison with the exact solution of the low frequency capacitance as shown in Fig. 4.11 reveals that the error can be substantial. The reasons for this is that we have ignored any charge variation in the semicondcutor. We will therefore now derive the exact flatband capacitance.

To derive the actual flatband capacitance we first linearize Poisson's equatiion. Since the potential across the semiconuctor at flatband is zero, we expect the potential to be small as we vary the gate voltage around the flatband voltage. Poisson's equation can then be simplified to :

$$\frac{d^2\phi}{dx^2} = \frac{q}{\varepsilon_s}\left(N_a^+ - p\right) = \frac{qN_a}{\varepsilon_s}\left(1 - e^{-\frac{\phi}{V_t}}\right) \cong \frac{qN_a}{\varepsilon_s}\frac{\phi}{V_t} \qquad \text{... (4.23)}$$

Charge due to ionized donors or electrons were ignored, since neither are present in a p-type semiconductor around flatband. The linearization is obtained by replacing the exponential function by the first two terms of its Tailor series expansion. The solution of this equation is:

$$\phi = \phi_s e^{-\frac{x}{L_D}}, \text{ with } L_D = \sqrt{\frac{\varepsilon_s V_t}{q N_a}} \qquad \text{... (4.24)}$$

Where f_s is the potential at the surface of the semiconductor and L_D is the Debye length. The solution of the potential enables the derivation of the capacitance of the semiconductor under flatband conditions, or :

$$C_{s,FB} = \frac{dQ_s}{d\phi_s} = \frac{d\left(\varepsilon_s \frac{\phi_s}{L_D}\right)}{d\phi_s} = \frac{\varepsilon_s}{L_D} \qquad \text{... (4.25)}$$

The flatband capacitance of the MOS structure at flatband is obtained by calculating the series connection of the oxide capacitance and the capacitance of the semiconductor, yielding :

$$C_{FB} = \frac{1}{\frac{1}{C_{ox}} + \frac{L_D}{\varepsilon_s}} \qquad \text{... (4.26)}$$

Example. *Calculate the oxide capacitance, the flatband capacitance and the high frequency capacitance in inversion of a silicon nMOS capacitor with a substrate doping $N_a = 10^{17}$ cm^{-3}, a 20 nm thick oxide ($e_{ox} = 3.9\ e_0$) and an aluminium gate ($F_M = 4.1$ V).*

Solution. The oxide capacitance equals :

$$C_{ox} = \frac{\varepsilon_{ox}}{t_{ox}} = \frac{3.9 \times 8.85 \times 10^{-14}}{2 \times 10^{-6}} = 173 \text{ nF/cm}^2.$$

The flatband capacitance equals :

$$C_{FB} = \frac{1}{\frac{1}{C_{ox}} + \frac{L_D}{\varepsilon_s}}$$

$$= \frac{1}{\frac{1}{173 \times 10^{-9}} + \frac{1.3 \times 10^{-6}}{11.9 \times 8.85 \times 10^{-14}}} = 142 \text{ nF/cm}^2.$$

where the Debye length is obtained from :

$$L_D = \sqrt{\frac{\varepsilon_s V_t}{q N_a}} = \sqrt{\frac{11.9 \times 8.85 \times 10^{-14} \times 0.0259}{1.6 \times 10^{-19} \times 10^{17}}} = 13 \text{ nm}$$

The high frequency capacitance in inversion equals :

$$C_{HF,inv} = \frac{1}{\frac{1}{C_{ox}} + \frac{x_{d,T}}{\varepsilon_s}}$$

$$= \frac{1}{\dfrac{1}{173\times10^{-9}} + \dfrac{1.05\times10^{-5}}{11.9\times8.85\times10^{-14}}} = 63 \text{ nF/cm}^2$$

and the depletion layer width at threshold equals :

$$x_{d,\text{T}} = \sqrt{\frac{2\varepsilon_s(2\phi_\text{F})}{qN_a}}$$

$$= \sqrt{\frac{2\times11.9\times8.85\times10^{-14}\times2\times0.419}{1.6\times10^{-19}\times10^{17}}} = 105 \text{ nm.}$$

The bulk potential, f_F, was already calculated in example 6.1.

4.3.4.3 Deep depletion capacitance

Deep depletion occurs in an MOS capacitor when measuring the high-frequency capacitance while sweeping the gate voltage "quickly". Quickly means that the gate voltage must be changed fast enough so that the structure is not in thermal equilibrium. One then observes that, when ramping the voltage from flatband to threshold and beyond, the inversion layer is not or only partially formed. This occurs since the generation of minority carriers cannot keep up with the amount needed to form the full inversion layer. The depletion laye therefore keeps increasing beyond its maximum thermal equilibrium value, $x_{d,\text{T}}$ resulting in a capacitance which further decreases with voltage.

The time required to reach thermal equilibrium can be estimated by taking the ratio of the total charge in the inversion layer to the thermal generation rate of minority carriers. A complete analysis should include both the surface generation rate as well as generation in the depletion layer and the quasi-neutral region. A good approximation is obtained by considering only the generation rate in the depletion region and the quasi-neutral region. This yields the following equation :

$$\text{time} = \frac{|Q_{inv}|}{qG(x_{d,dd}+L_n)} = \frac{C_{ox}(V_G - V_T)}{\dfrac{qn_i}{2\tau}\sqrt{\dfrac{2\varepsilon_s\phi_{s,dd}}{qN_a}} + (\mu_n V_t \tau)} \quad \text{... (4.27)}$$

where the generation in the depletion layer was assumed to be constant. For materials with a long diffusion length one can also ignore the generation in the depletion region. The rate of change required to observe deep depletion is then obtained from :

$$\frac{qV_G}{dt} > \frac{qn_i}{2C_{ox}}\sqrt{\frac{\mu_n V_t}{\tau}} \quad \text{... (4.28)}$$

This equation predicts that deep depletion is less likely at higher ambient temperature, since the intrinsic carrier density n_i increases exponentially with temperature. The intrinsic density also decreases exponentially with the energy bandgap. Therefore, MOS structures made with wide bandgap materials (for instance 6H-SiC for which E_g = 3 eV), have an extremely pronounced deep depletion effect.

In silicon MOS capacitors, one finds that the occurrence of deep depletion can be linked to the minority carrier lifetime. Structures with a long (0.1 ms) lifetime require a few seconds to reach thermal equilibrium, which results in a pronounced deep depletion effect at room temperature. Structures witha short (1 ms) lifetime do not show this effect

Carrier generation due to light will increase the generation rate beyond the ghermal generation rate and therefore reduces the time needed to reach equilibrium. Deep depletion measurements are therefore done in the dark.

4.3.4.4 *Experimental results and comparison with theory*

As an example, we present the measured low frequency (quasi-static) and high frequency capacitance- voltage curves of an MOS capacitor. The capacitance was measured in the presence of ambient light as well as in the dark as explained in he caption of Fig. 4.12

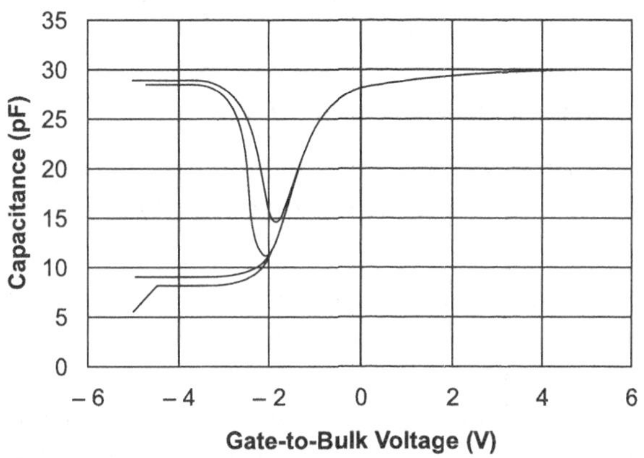

Fig. 4.12 *Low frequency (quasi-static) and high frequency capacitance measurement of a pMOS capacitor. Shown are, from top to bottom, the low frequency capacitance measured in the presence of ambient light (top curve), the low frequency capacitance measured in the dark, the high frequency capacitance measured in the presence of ambient light and the high frequency capacitance measured in the dark (bottom curve). All curves were measured from left to right. The MOS parameters are* $N_a - N_d = 4 \times 10^{15}$ *cm*$^{-3}$ *and* t_{ox} = *80 nm. The device area is 0.0007 cm*^{2}

Figure 4.12. illustrates some of the issues when measuring the capacitance of an MOS capacitance. First, one should measure the devices in the dark. The presence of light causes carrier generation in the semiconductor, which affects the measured capacitance. In addition, one must avoid the deep depletion effects such as the initial linearly varying capacitance of the high frequency capacitance measured in the dark on the above Figure (bottom curve). The larger the carrier lifetime, the lower the voltage is to be changed to avoid deep depletion.

The measured low frequency capacitance is compared to the theoretical value in Fig. 4.13. The high frequency capacitance measured in the presence of light is also shown on the figure. The Figure 4.13 illustrates the agreement between experiment and theory.

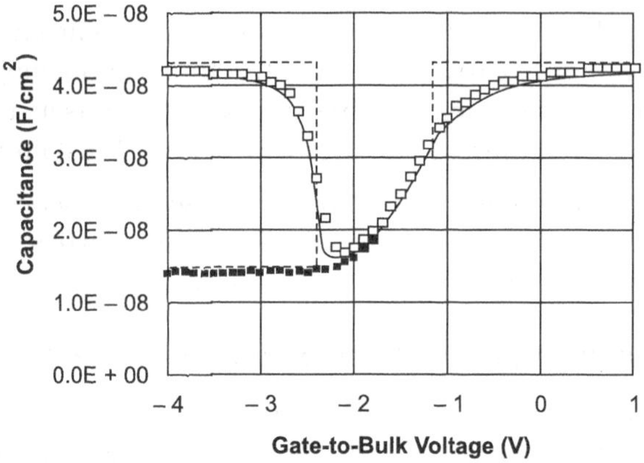

Fig. 4.13 *Comparison of the theoretical low frequency capacitance (solid line) and the experimental data (open squares) obtained in the dark. Fitting parameters are* $N_a - N_d = 3.95 \times 10^{15}\ cm^{-3}$ *and* $t_{ox} = 80\ nm$.

4.3.4.5 Non-Ideal effects in MOS capacitors

Non-ideal effects in MOS capacitors include fixed charge, mobile charge and charge in surface states. Performing a capacitance-voltage measurement can identify all three types of charge.

Fixed charge in the oxide simply shifts the measured curve. A positive fixed charge at the oxide-semiconductor interface shits the flatband voltage by an amount, which equals the charge divided by the oxide capacitance. he shift reduces linearly as one reduces the position of the charge relative to the gate electrode and becomes zero if the charge is located that the metal-oxide interface. A fixed charge is caused by ions, which are incorporated in the oxide during growth or deposition.

The flatband voltage shift due to **mobile charge** is described by the same equation as that due to fixed charge. However, the measured curves differ since a positive gate voltage causes any negative mobile charge to move away from the gate electrode, while a negative voltage attracts the charge towards the gate. This causes the curve to shift towards the applied voltage. One can recognize mobile charge by the hysteresis in the high frequency capacitance curve when sweeping the gate voltage back and fofth. Sodium ions incorporated in the oxide of silicon MOS capacitors are known to yield mobile charges. It is because of the high densitivity of MOS structures to a variety of impurities that the industry carefully controls the purity of the water and the chemicals used.

Charge due to electrons occupying **surface states** also yields a shift in flatband voltage. However as the applied voltage is varied, the Fermi energy at the oxide-semiconductor interface changes also and affects the occupancy of the surface states. The interface states cause the transition in the capacitance measurement to be less abrupt. The combination of the low frequency and high frequency capacitance allows a calculation of the surface state density. This method provides the surface state density over a limited (but highly relevant) range of energies within the bandgap. Measurements of n-type and p-type capacitors at different temperatures provide the surface state density throughout the bandgap.

4.3.5 Derivation of the Exact Solution

We now derive the exact solution of the MOS capacitor. Whereas most of the derivation is applicable for *n*-type and *p*-type substrates, the equations are written in a form, which is more convenient for *p*-type substrates.

The total charge density, *r*, in the semiconductor is given by :

$$\rho = q\left(p + N_d^+ - n - N_a^-\right) \qquad \text{... (4.29)}$$

Under thermal equilibrium, the hole and electron densities, *p* and *n*, can be expressed as a function of the potential, *f*, and a reference potential, f_F.

$$p = n_i \exp\left(\frac{\phi_F - \phi(x)}{V_t}\right) \qquad \text{... (4.30)}$$

$$n = n_i \exp\left(\frac{\phi(x) - \phi_F}{V_t}\right) \qquad \text{... (4.31)}$$

Far away from the oxide-semiconductor interface, the charge density is zero and we define the potential, *f*, to be zero there also, so that

$$N_d^+ - N_a^- = -2n_i \sinh\left(\frac{\phi_F}{V_t}\right) \qquad \text{... (4.32)}$$

Poisson's equation then takes the following form :

$$\frac{d^2\phi}{dx^2} = \frac{2qn_i}{\varepsilon_s}\left\{\sinh\left(\frac{\phi - \phi_F}{V_t}\right) + \sinh\left(\frac{\phi_F}{V_t}\right)\right\} \qquad \text{... (4.33)}$$

Multiplying both sides of the equation with $2\,\frac{df}{dx}$ and integrating while replacing $-\frac{df}{dx}$ by the electric field ε, one obtains :

$$\varepsilon(\phi) = \text{sign}(\phi)\sqrt{\frac{4qn_i V_t}{\varepsilon_s}\left(\cosh\left(\frac{\phi - \phi_F}{V_t}\right) + \frac{\phi}{V_t}\sinh\left(\frac{\phi_F}{V_t}\right) + K\right)} \qquad \text{... (4.34)}$$

The constant K can be determined from the boundary condition at x = ¥, where f = ε = 0 yielding :

$$K = -\cosh\left(\frac{\phi_F}{V_t}\right) \qquad \text{... (4.35)}$$

The electric field has the same sign as the potential as indicated by the sign function.

The relation between the field and the potential at the surface under thermal equilibrium is then :

$$\varepsilon_{s,eq} = 2\,\text{sign}\,(\phi_s)\sqrt{\frac{qn_i V_t}{\varepsilon_s}\left\{\cosh\left(\frac{\phi_s - \phi_F}{V_t}\right) + \frac{\phi_s}{V_t}\sinh\left(\frac{\phi_F}{V_t}\right) - \cosh\left(\frac{\phi_F}{V_t}\right)\right\}} \quad \text{... (4.36)}$$

The gate voltage equals the sum of the flatband voltage, the voltage across the oxide and the potential across the semiconductor :

$$V_G = V_{FB} + \phi_s + V_{ox}, \text{ with } V_{ox} = t_{ox}\,\varepsilon_{s,eq}\,(\phi_s)\,\frac{\varepsilon_s}{\varepsilon_{ox}} \quad \text{... (4.37)}$$

4.3.5.1 Low frequency capacitance

The low frequency capacitance of the MOS structure per unit area can now be calculated from :

$$C_{LF} = \left|\frac{dQ_s}{dV_G}\right| = \varepsilon_s \frac{d\varepsilon_{s,eq.}}{d\left(\phi_s + t_{ox}\,\varepsilon_{s,eq.}\,(\phi_s)\dfrac{\varepsilon_s}{\varepsilon_{ox}}\right)} = \frac{1}{\dfrac{1}{C_{ox}} + \dfrac{1}{C_{s,LF}}} \quad \text{... (4.38)}$$

where $C_{os} = \dfrac{\varepsilon_{ox}}{t_{ox}}$ and

$$C_{s,LF} = \varepsilon_s \frac{d\varepsilon_{s,eq.}}{d\phi_s}$$

$$= \varepsilon_s \sqrt{\frac{qn_i}{\varepsilon_s V_t}} \frac{\left|\sinh\left(\dfrac{\phi_s - \phi_F}{V_t}\right) + \sinh\left(\dfrac{\phi_F}{V_t}\right)\right|}{\sqrt{\cosh\left(\dfrac{\phi_s - \phi_F}{V_t}\right) + \dfrac{\phi_s}{V_t}\sinh\left(\dfrac{\phi_F}{V_t}\right) - \cosh\left(\dfrac{\phi_F}{V_t}\right)}} \quad \text{... (4.39)}$$

$$C_{s,LF} = 2\frac{qn_i}{\varepsilon_{s,eq}}\left[\sinh\left(\frac{\phi_s - \phi_F}{V_t}\right) + \sinh\left(\frac{\phi_F}{V_t}\right)\right] \quad \text{... (4.40)}$$

This result is often referred to as the low frequency capacitance of a MOS capacitor since we calculated the change in charge between two equilibrium situations. The result can be interpreted as a series connection of the oxide capacitance and the low frequency capacitance of the semiconductor $C_{s,\,LF}$. By starting rom a series of values for f_s, one can use the above equations to first calculate the electric field, the gate voltage and the capacitance. This enables to plot the low frequency capacitance as a function of the gate voltage as shown in Fig. 4.14.

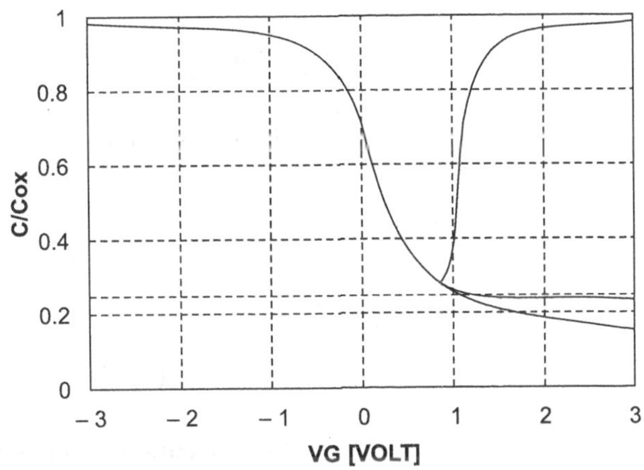

Fig. 4.14 *Capacitance versus voltage for a MOS structure with $N_a = 10^{15}$ cm^{-3} and t_{ox} = 0.1 mm. The curves from top to bottom are : The low frequency capacitance, the approximate high frequency capacitance, the exact high frequency capacitance and the capacitance under deep depletion conditions. The dotted line indicates C_{min}, the capacitance at the onset of strong inversion as calculated using the full depletion approximation*

Under flatband conditions, where $0 = | f_s | << V_t$, the capacitance reduces to

$$C_{FB} = \frac{1}{\dfrac{1}{C_{ox}} + \dfrac{L_D}{\varepsilon_s}} \quad \text{... (4.41)}$$

where L_D is the extrinsic Debye length in the semiconductor with doping $| N_a - N_d |$:

$$L_D = \sqrt{\frac{\varepsilon_s V_t}{q|N_a - N_d|}} \quad \text{... (4.42)}$$

4.3.5.2 Deep depletion capacitance

If the gate voltage is changed faster than electrons can be generated at the oxide-semiconductor interface to obtain the equilibrium density, no inversion layer is generated. Inthis case the gate voltage will cause the depletion layer in the semiconductor to exceed the maximum depletion layer width as defined at the onset of strong inversion. A typical measurement starts from an equilibrium situation where no inversion layer is present and the gate voltage is swept rapidly while creating a depletion layer in the semiconductor. The capacitance is measured as the change in charge flowing into the structure for a given voltage change. For a p-type substrate, this situation can be modeled by eliminating the charge term due to electrons in Poisson's equation :

$$\frac{d^2\phi}{dx^2} = \frac{qn_i}{\varepsilon_s}\left\{2\sinh\left(\frac{\phi_F}{V_t}\right) - \exp\left(\frac{\phi_F - \phi}{V_t}\right)\right\} \quad \text{... (4.42)}$$

Using the same procedure as above the relation between surface field and surface potential can be found :

$$\varepsilon_{s,dd} = \text{sifn}(\phi_s)\sqrt{\frac{2qn_i V_t}{\varepsilon_s}\left\{2\frac{\phi_s}{V_t}\sinh\left(\frac{\phi_F}{V_t}\right) + \exp\left(\frac{\phi_F}{V_t}\right)\left(\exp\left(-\frac{\phi_s}{V_t}\right) - 1\right)\right\}} \quad \text{... (4.43)}$$

and the capacitance of the semiconductor becomes :

$$C_{s,dd} = \left|\frac{qn_i}{\varepsilon_{s,dd}}\left[2\sinh\left(\frac{\phi_F}{V_t}\right) - \exp\left(\frac{\phi_F - \phi_s}{V_t}\right)\right]\right| \left|\frac{qn_i}{\varepsilon_{s,dd}}\left[2\sinh\left(\frac{\phi_F}{V_t}\right) - \exp\left(\frac{\phi_F - \phi_s}{V_t}\right)\right]\right|$$

The corresponding gate voltage is :

$$V_G = V_{FB} + \phi_s + V_{ox}, \text{ with } V_{ox} = t_{ox}\,\varepsilon_{s,dd}(\phi_s)\,\frac{\varepsilon_s}{\varepsilon_{ox}} \quad \text{... (4.44)}$$

Using a similar procedure as for the low frequency capacitance we can also plot the capacitance under deep depletion conditions.

4.3.5.3 High frequency capacitance

The high frequency capacitance of an MOS capacitor is measured by applying a small ac voltage in addition to the dc gate voltage. The capacitance is defined as the ratio of the out-of-phase component of the ac current divided by the amplitude of the ac voltage times the radial frequency. An approximate expression can be obtained by ignoring the change in charge in the inversion layer yielding the expression for the capacitance under deep depletion conditions. However since the gate voltage is changed slowly while measuring the capacitance versus voltage, the gate voltage is calculated from the surface potential including the charge in the inversion layer under thermal equilibrium. The capacitance is then given by :

$$C_{s,HF} = \left|\frac{qn_i}{\varepsilon_{s,dd}}\left[2\sinh\left(\frac{\phi_F}{V_t}\right) - \exp\left(\frac{\phi_F - \phi_s}{V_t}\right)\right]\right| \quad \text{... (4.45)}$$

with the electric field, $\varepsilon_{s,dd}$, obtained under deep depletion conditions (4.5.16).

This is the same expression as for the capacitance under deep depletion conditions. However, the corresponding gate voltage is different, namely :

$$V_G = V_{FB} + \phi_s + V_{ox}, \text{ with } V_{ox} = t_{ox}\,\varepsilon_{s,eq}(\phi_s)\,\frac{\varepsilon_s}{\varepsilon_{ox}} \quad \text{... (4.46)}$$

where the electric field, $\varepsilon_{s,eq}$, is the thermal equilibrium field

The corresponding capacitance is also included together with the expected minimum capacitance based on the full depletion approximation, corrected for the thermal voltage:

$$\frac{1}{C_{min}} = \frac{1}{C_{ox}} + \sqrt{\frac{2(2\phi_F + V_t)}{qN_a\varepsilon_s}} \quad \text{... (4.47)}$$

It should be noted that this is only an approximate solution. The redistribution of the inversion layer charge with applied gate voltage is ignored even though it does affect the depletion layer width and with it the capacitance. This approximation therefore introduces an error which was found to be less than 6% at the onset of strong inversion and which increases almost linearly with increasing surface potential.

The exact expression for the high frequency capacitance used is :

$$C_{s,\,HF,\,exact} = \frac{qn_i \operatorname{sign}(\phi_s)}{\varepsilon_{s,\,eq}}$$

$$\times \left\{ \exp(U_F)\left[1 - \exp(-U_s)\right] + \exp(-U_F)\frac{\left[\exp(U_s) - 1\right]}{1 + \Delta} \right\} \quad \text{... (4.48)}$$

Where D for a p-type substrate is :

$$D = 0 \text{ for } \phi_s < 0 \text{ and } \phi_F > 0 \quad \text{... (4.49)}$$

$$\Delta = \frac{\dfrac{(\exp(U_s) - U_s - 1)}{F(U_s/U_F)}}{\displaystyle\int_0^{U_s} \frac{\exp(U_F)(1 - \exp(-\xi))(\exp(-\xi) - \xi - 1)}{2F^3(\xi/U_F)}\, d\xi} \text{ for } \phi_s > 0 \text{ and } \phi_F > 0 \quad \text{... (4.50)}$$

The expression with D = 0 for all possible surface potentials equals the low frequency capacitance. The function F is related to the equilibrium electric field by :

$$F\left(\frac{U}{U_F}\right) = \frac{\varepsilon_{eq} L_{D,\,i}}{2\sqrt{2}\, V_t} \quad \text{... (4.51)}$$

and the normalized parameters U, U_s and U_F are defined as :

$$U = \frac{\phi}{V_t},\ U_3 = \frac{\phi_s}{V_t},\ U_F = \frac{\phi_F}{V_t} \quad \text{... (4.52)}$$

Where the gate voltage is still given by :

$$V_G = V_{FB} + \phi_s + V_{ox}, \text{ with } V_{ox} = t_{ox}\, \varepsilon_{s,\,eq}(\phi_s)\, \frac{\varepsilon_s}{\varepsilon_{ox}} \quad \text{... (4.53)}$$

and the electric field, $\varepsilon_{s,\,eq}$, is the thermal equilibrium field at the surface.

As illustrated with Fig. 4.5.4, the high frequency capacitance at the onset of strong inversion ($f_s = 2f_F$) and beyond is found to be almost constant. Assuming $f_F >> V_t$ equation (4.5.19) becomes :

$$C_{s,\,HF} = \sqrt{\frac{qN_a \varepsilon_s}{4\phi_F}},\ \text{for } V_G - V_{FB} > 2\phi_F + \frac{\sqrt{4qN_a \varepsilon_s \phi_F}}{C_{ox}} \quad \text{... (4.54)}$$

This result could also be obtained by calculating the depletion region width in the semiconductor assuming the maximum potential at the surface to be $2f_F$ and using the full depletion approximation.

The low frequency capacitance (equation (4.5.12)) at $f_s = 2f_F$, assuming $f_F >> V_t$ is then:

$$C_{s,LF} = \sqrt{\frac{2qN_a\varepsilon_s}{\phi_F}}, \text{ for } V_G - V_{FB} = 2\phi_F + \frac{\sqrt{4qN_a\varepsilon_s\phi_F}}{C_{ox}} \quad \text{... (4.55)}$$

also yielding the following relation between both :

$$C_{s,LF} = \sqrt{8}\,C_{s,HF}, \text{ for } \phi_s = 2\phi_F \quad \text{... (4.56)}$$

4.4 p-MOS AND GENERAL EQUATIONS

Throughout this text we have chosen to use nMOS capacitors are the preferred example. Several equations have only been provided for nMOS as those for pMOS are very similar. In this section we provide equations specifically for pMOS capacitors as well as a set of equations that could be applied to either type.

4.4.1 pMOS Equations

pMOS capacitors have an n-type substrate, a positive charge in the depletion layer and a positive charge in the inversion layer. Since the Fermi energy is a distance $q\phi_F$/sub > above the midgap energy level, the work function difference is given by :

$$\Phi_{MS} = \Phi_M - \Phi_S = \Phi_M - \left(\chi + \frac{E_g}{2q} - \phi_F\right), n\text{-substrate} \quad \text{... (4.57)}$$

with $$|\phi_F| = V_t \ln\frac{N_d}{n_i}, n\text{-substrate} \quad \text{... (4.58)}$$

The expression for the depletion layer width is similar to that of nMOS capacitors, namely :

$$x_{d,T} = \sqrt{\frac{2\varepsilon_s 2|\phi_F|}{qN_d}}, n\text{-substrate} \quad \text{... (4.59)}$$

while the threshold is typically negative due to the positive charge in the depletion layer width.

$$V_T = V_{FB} - 2|\phi_F| - \frac{\sqrt{2\varepsilon_s 2|\phi_F|qN_d}}{C_{ox}}, n\text{-substrate} \quad \text{... (4.60)}$$

Note that the relation between the flatband voltage and the workfunction difference still applies:

$$V_{FB} = \Phi_{MS} - \frac{Q_i}{C_{ox}} - \frac{1}{\varepsilon_{ox}}\int_0^{t_{ox}} \rho_{ox}(x)\,x\,dx, n\text{-substrate} \quad \text{... (4.61)}$$

4.4.2 General Equations

General equations, which are valid for n-MOS and p-MOS capacitors are provided below. The type is directly linked to the net doping density of the substrate, $N_a - N_d$, which is positive for a p-type substrate (n-MOS capacitor) and negative for an n-type substrate (p-MOS capacitor). The workfunction difference is then given by :

$$\Phi_{MS} = \Phi_M - \Phi_S = \Phi_M - \left(\chi + \frac{E_c - E_i}{2q} + \phi_F\right), n\text{-substrate} \qquad \text{... (4.62)}$$

where the built-in potential is positive for p-type substrates and negative for n-type substrates and is given by :

$$\phi_F = V_t \ln \frac{p}{n_i} = -V_t \ln \frac{n}{n_i}, n\text{-substrate} \qquad \text{... (4.63)}$$

The depletion layer width at threshold is then :

$$x_{d,T} = \sqrt{\frac{2\varepsilon_2 2\phi_F}{q(N_a - N_d)}} \qquad \text{... (4.64)}$$

and the threshold voltage is given by :

$$V_T = V_{FB} + 2\phi_F + \frac{q(N_a - N_d)}{C_{ox}}\sqrt{\frac{4\varepsilon_s \phi_F}{q(N_a - N_d)}} \qquad \text{... (4.65)}$$

These equations are of interest when parameters of n-MOS as well as p-MOS capacitors are to be calculated. The equations eliminate the problem of the variable signs and possible mistakes and confusion, at the expense of the added complexity.

In a MOSFET structure it is possible to apply a voltage, V_C, to the channel relative to the voltage at the bulk contact to the substrate, V_B. This affects the width of the depletion layer width at threshold:

$$x_{d,T} = \sqrt{\frac{2\varepsilon_s (2\phi_F + V_C - V_B)}{q(N_a - N_d)}} \qquad \text{... (4.66)}$$

as well as the threshold voltage itself :

$$V_T = V_{FB} + V_C + 2\phi_F + \frac{q(N_a - N_d)}{C_{ox}}\sqrt{\frac{2\varepsilon_s (2\phi_F + V_C - V_B)}{q(N_a - N_d)}} \qquad \text{... (4.67)}$$

QUESTIONS

- Drawn an MOS flatband diagram. Indicate the workfunction of the metal and the semiconductor as well as the flatband voltage. Draw it approximately to scale using $\Phi_M = 4.1$ V, $\chi = 4.05$ V, $E_g = 1.12$ eV (silicon) and $N_a = 10^{16}$ cm^{-3}.
- Derive the metal-semiconductor workfunction for n-type and p-type poly-silicon gate structures.
- Explain why the flatband voltage depends on the charge in the oxide or at the oxide-semiconductor interface.
- Name the three bias regimes of an MOS capacitor and explain what happens in the semiconductor in each of these bias modes.
- What is the basic assumption regarding the charge in the inversion layer ?
- What are the assumptions of the MOS capacitor analysis ?
- What is the difference between the high frequency and quasi-static capacitance ?
- Why is he high frequency capacitance constant in inversion ?
- Why does the flatband capacitance not equal the oxide capacitance ?
- What is deep depletion ?
- Why does light illumination affect the capacitance of an MOS structure ?
- Name the non-ideal effects in MOS capacitors. What causes them and how do they affect the MOS characteristics ?

5

MOS Field-Effect-Transistors

5.1 INTRODUCTION

The n-type Metal-Oxide-Semiconductor Field-Effect-Transistor (nMOSFET) consists of a source and a drain, two highly conducting n-type semiconductor regions, which are isolated from the p-type substrate by reversed-biased p-n diodes. A metal or poly-crystalline gate covers the region between source and drain. The gate is separated from the semiconductor by the gate oxide. The basic structure of an n-type MOSFET and the corresponding circuit symbol are shown in Fig. 5.1.

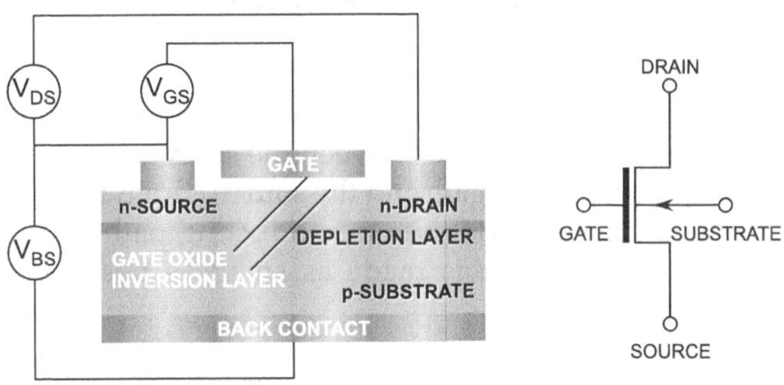

Fig. 5.1 *Cross-section and circuit symbol of an n-type Metal-Oxide-Semiconductor-Field-Effect-Transistor (MOSFET)*

As can be seen on the Figure, the source and drain regions are identical. It is the applied voltages, which determine which n-type region provides the electrons and becomes the source, while the other n-type region receives the electrons and becomes the drain. The voltages applied to the drain and gate electrode as well as to the substrate, by means of a back contact, are referred to the source potential, as also indicated in Fig. 5.1.

A conceptually similar structure was proposed and patented independently by Lilienfeld and Heil in 1930, but the MOSFET was not successfully demonstrated until 1960. The main technological problem was the control and reduction of the surface states at the oxide-semiconductor interface.

Initially, it was only possible to deplete an existing *n*-type channel by applying a negative voltage to the gate. Such devices have a conducting channel between source and drain even when no gate voltage is applied. They are called "depletion-mode" devices.

A reduction of the surface states enabled the fabrication of devices, which do not have a conducting channel unless a positive voltage is applied. Such devices are referred to as "enhancement-mode" devices. The electrons at the oxide-semiconductor interface are concentrated in a thin (~ 10 nm thick) "inversion" layer. By now, most MOSFETs are "enhancement-mode" devices.

While a minimum requirement for amplification of electrical signals is power gain, one finds that a device with both voltage and curent gain is a highly desirable circuit element. The MOSFET provides current and voltage gain yielding an output current into an external load, which exceeds the input current, and an output voltage across that external load which exceeds the input voltage.

The current gain capability of a Field-Effect-Transistor (FET) is easily explained by the fact that no gate current is required to maintain the inversion layer and the resulting current between drain and source. The device has therefore an infinite current gain in dc. The current gain is inversely proportional to the signal frequency, reaching unity current gain at the transit frequency.

The voltage gain of the MOSFET is caused by the current saturation at higher drain-source voltages, so that a small drain-current variation can cause a large drain-voltage variation.

5.2 STRUCTURE AND PRINCIPLE OF OPERATION

A top view of the same MOSFET is shown in Fig. 5.2, where the gate length, L, and gate width, W, are identified. Note that the gate length does not equal the physical dimension of the gate, but rather the distance between the source and drain regions underneath the gate. The overlap between the gate and the source/drain region is required to ensure that the inversion layer forms a continuous conducting path between the source and drain region. Typically this overlap is made as small as possible in order to minimize its parasitic capacitance.

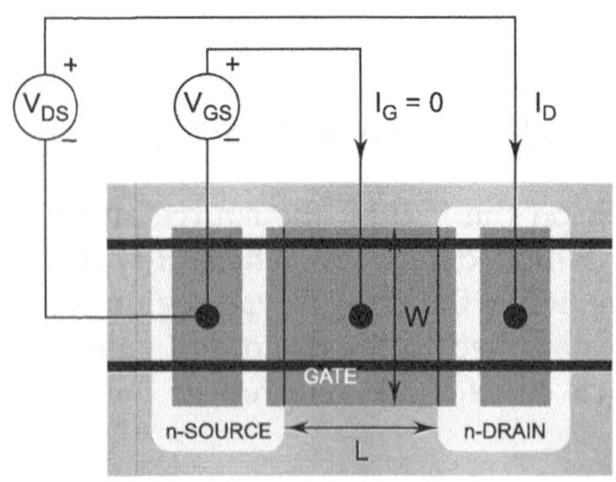

Fig. 5.2 *Top view of an n-type Metal-Oxide-Semiconductor-Field-Effect-Transistor (MOSFET)*

The voltage applied to the gate controls the flow of electrons from the source to the drain. A positive voltage applied to the gate attracts electrons to the interface between the gate dielectric and the semiconductor. These electrons form a conducting channel called the inversion layer. No gate current is required to maintain the inversion layer at the interface since the gate oxide blocks any carrier flow. The net result is that the applied gate voltage controls the current between drain and source.

The typical current versus voltage (I-V) characteristics of a MOSFET are shown in Fig. 5.3.

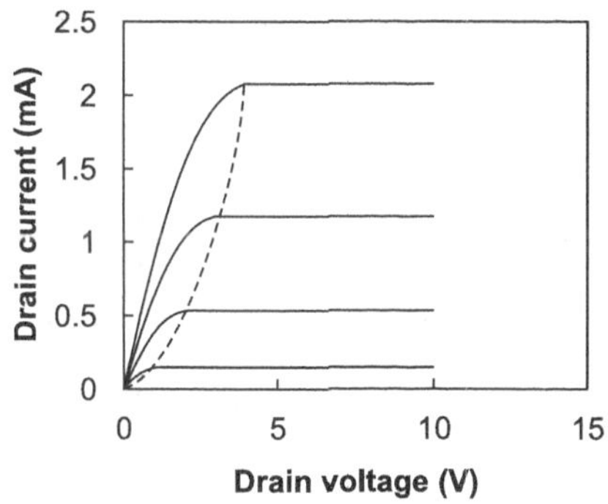

Fig. 5.3 *I-V characteristics of an n-type MOSFET with* $V_G = 5$ *V (top curve), 4 V, 3 V and 2 V (bottom curve)*

Note: We will primarily consider the *n*-type or *n*-channel MOSFET in this chapter. This type of MOSFET is fabricated on a *p*-type semiconductor substrate. The complementary MOSFET is the *p*-type or *p*-channel MOSFET. The *p*-type MOSFET contains *p*-type source and drain regions in an *n*-type substrate. The inversion layer is formed when holes are attracted to the interface by a negative gate voltage. While the holes still flow from source to drain, they result in a negative drain current. CMOS circuits require both *n*-type and *p*-type MOSFETs.

5.3 MOSFET ANALYSIS

In this section, we present three different models for the MOSFET, the linear model, the quadratic model and the variable depletion layer model. The linear model correctly predicts the MOSFET behaviour for small drain-source voltages, where the MOSFET acts as a variable resistor. The quadratic model includes the voltage variation along the channel between source and drain. This model is most commonly used despite the fact that the variation of the depletion layer charge is ignored. The variable depletion layer model is more complex as it includes the variation of the depletion layer along the channel.

5.3.1 The Linear Model

The linear model describes the behaviour of a MOSFET biased with a small drain-to-source voltage. As the name suggests, the linear model, describes the MOSFET acting as a linear device. More specifically, it can be modeled as a linear resistor whose resistance is modulated by the gate-to-source voltage. In this regime, the MOSFET can be used as a switch for analog and digital signals or as an analog multiplier.

The general expression for the drain current equals the total charge in the inversion layer divided by the time the carriers need to flow from the source to the drain:

$$I_D = -\frac{Q_{inv}\,WL}{t_r} \qquad \text{... (5.1)}$$

where Q_{inv} is the inversion layer charge per unit area, W is the gate width, L is the gate length and t_r is the transit time. If the velocity of the carriers is constant between source and drain, the transit time equals:

$$t_r = \frac{L}{v} \qquad \text{... (5.2)}$$

where the velocity, v, equals the product of the mobility and the electric field:

$$v = \mu\varepsilon = \mu\frac{V_{DS}}{L} \qquad \text{... (5.3)}$$

The constant velocity also implies a constant electric field so that the field equals the drain-source voltage divided by the gate length. This leads to the following expression for the drain current:

$$I_D = -\mu\,Q_{inv}\cdot\frac{W}{L}.V_{DS} \qquad \text{... (5.4)}$$

Where

$$Q_{inv} = -C_{ox}\,(V_{GS}-V_T),\ \text{for } V_{GS} > V_T \qquad \text{... (5.5)}$$

The inversion layer charge is zero if the gate voltage is lower than the threshold voltage. Replacing the inversion layer charge density in the expression for the drain current yields the linear model:

$$I_D = \mu C_{ox}\frac{W}{L}(V_{GS} - V_T)\,V_{DS},\ \text{for } |\,V_{DS}\,| << (V_{GS}-V_T) \qquad \text{... (5.6)}$$

Note that the capacitance in the above equations is the gate oxide capacitance per unit area. Also note that the drain current is zero if the gate-to-source voltage is less than the threshold voltage. The linear model is only valid if the drain-to-source voltage is much smaller than the gate-to-source voltage minus the threshold voltage. This insures that the velocity, the electric field and the inversion layer charge density is indeed constant between the source and the drain.

An example of the linear current-versus-voltage (I-V) characteristics of a MOSFET is shown in Fig. 5.4.

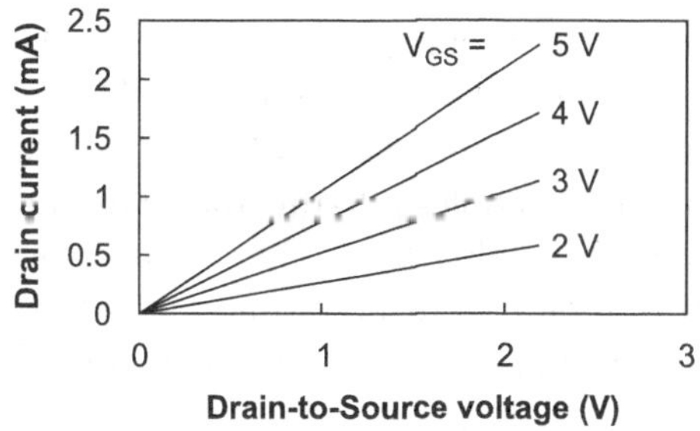

Fig. 5.4 *Linear I-V characteristics of a MOSFET with $V_T = 1$ V, ($\mu_n = 300$ cm^2/V-s, W/ L = 5 and $t_{ox} = 20$ nm).*

The Fig. 5.4 illustrates the behaviour of the device in the linear regime : While there is no drain current if the gate voltage is less than the threshold voltage, the current increases with gate voltage once it is larger than the threshold voltage. The slope of the curves equals the conductance of the device, which increases linearly with the applied gate voltage. The figure therefore illustrates the use of a MOSFET as a voltage-controlled resistor.

5.3.2 The Quadratic Model

The quadratic model uses the same assumptions as the linear model. However, this model allows the inversion layer charge to vary between the source and the drain.

The derivation is based on the fact that the current is continuous throughout the channel. The current is also related to the local channel voltage, > I > V_C.

We now consider a small section within the device with width *dy* > / I > and channel voltage $V_C + V_S$. The linear model is described by equation:

$$I_D = \mu C_{ox} \frac{W}{dy} (V_G - V_S - V_C - V_T)\, dV_C \qquad \text{... (5.7)}$$

where the drain-source voltage is replaced by the channel voltage. Both sides of the equation can be integrated from the source to the drain, so that *y* varies from 0 to the gate length, L, and the channel voltage V_C varies from 0 to the drain-source voltage, V_{DS}.

$$\int_0^L I_D\, dy = \mu C_{ox} W \int_0^{V_{DS}} (V_G - V_S - V_C - V_T)\, dV_C \qquad \text{... (5.8)}$$

The drain current, I_D, is constant so that integration results in:

$$I_D = \mu C_{ox} \frac{W}{L} \left[(V_{GS} - V_T) V_{DS} - \frac{V_{DS}^2}{2} \right], \text{ for } V_{DS} < V_{GS} - V_T \qquad \text{... (5.9)}$$

The drain current first increases linearly with the applied drain-to-source voltage, but then reaches a maximum value. According to the above equation the current would even decrease and eventually become negative. The charge density at the drain end of the channel is zero at that maximum and changes sign as the drain current decreases. As explained in section 4.2, the charge in the inversion layer does go to zero and reverses its sign as holes are accumulated at the interface. However, these holes cannot contribute to the drain current since the reversed-biased *p-n* diode between the drain and the substrate blocks any flow of holes into the drain. Instead the current reaches its maximum value and maintains that value for higher drain-to-source voltages. A depletion layer located at the drain end of the gate accommodates the additional drain-to-source voltage. This behaviour is referred to as *drain current saturation.*

Drain current saturation therefore occurs when the drain-to-source voltage equals the gate-to-source voltage minus the threshold voltage. The value of the saturated drain current, $I_{D,\,sat.}$, is then given by the following equation:

$$I_{D,sat} = \mu C_{ox} \frac{W}{L} \frac{(V_{GS} - V_T)^2}{2}, \text{ for } V_{DS} > V_{GS} - V_T \qquad \text{... (5.10)}$$

The quadratic model explains the typical current-voltage characteristics of a MOSFET, which are normally plotted for different gate-to-source voltages. An example is shown in Fig. 5.5. The saturation occurs to the right of the dotted line which is given by $I_D = \mu C_{ox} \frac{W}{L} V_{DS}^2$.

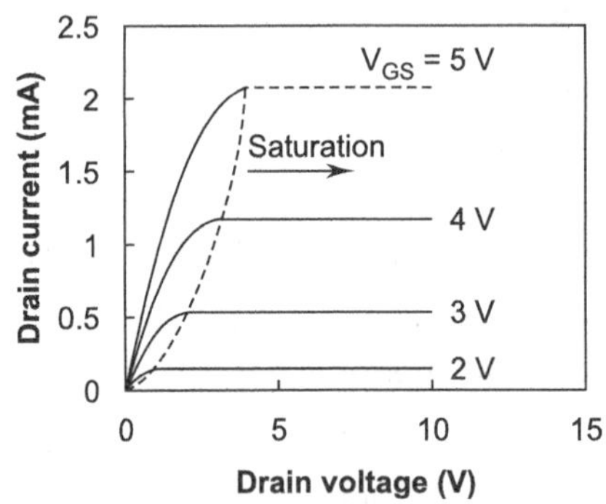

Fig. 5.5 *Current-Voltage characteristics of an n-type MOSFET as obtained with the quadratic model. The dotted line separates the quadratic region of operation on the left from the saturation region on the right.*

The drain current is still zero if the gate voltage is less than the threshold voltage.

$$I_D = 0, \qquad \text{for } V_{GS} < V_T \qquad \text{... (5.11)}$$

For negative drain-source voltages, the transistor is in the quadratic regime and is described by equation (5.9). However, it is possible to forward bias the drain-bulk *p-n* junction. A complete circuit model should therefore also include the *p-n* diodes between the source, the drain and the substrate.

We now use the quadratic model used to calculate some of the small signal parameters, namely the transconductance, g_m and the output conductance, g_d.

The transconductance quantifies the drain current variation with a gate-source voltage variation while keeping the drain-source voltage constant, or:

$$g_m \stackrel{\Delta}{=} \left. \frac{\partial I_D}{\partial V_{GS}} \right|_{V_{DS}} \qquad \text{... (5.12)}$$

The transconductance in the quadratic region is given by:

$$g_{m,quad} = \mu C_{ox} \frac{W}{L} V_{DS} \qquad \text{... (5.13)}$$

which is proportional to the drain-source voltage for $V_{DS} < V_{GS} - V_T$. In saturation, the transconductance is constant and equals:

$$g_{m,sat} = \mu C_{ox} \frac{W}{L}(V_{GS} - V_T) \qquad \text{... (5.14)}$$

The output conductance quantifies the drain current variation with a drain-source voltage variation while keeping the gate-source voltage constant, or:

$$g_d \triangleq \left. \frac{\partial I_D}{\partial V_{DS}} \right|_{V_{GS}} \qquad \text{... (5.15)}$$

The output conductance in the quadratic region decreases with increasing drain-source voltage:

$$g_{d,quad} = \mu C_{ox} \frac{W}{L}(V_{GS} - V_T - V_{DS}) \qquad \text{... (5.16)}$$

and becomes zero as the device is operated in the saturated region:

$$g_{d,sat} = 0 \qquad \text{... (5.17)}$$

Example. 5.1. *Calculate the drain current of a silicon nMOSFET with $V_T = 1$ V, $W = 10$ μm, $L = 1$ μm and $t_{ox} = 20$ nm. The device is biased with $V_{GS} = 3$ V and $V_{DS} = 5$ V. Use the quadratic model, a surface mobility of 300 cm^2/V-s and set $V_{BS} = 0$ V.*

Also calculate the transconductance at $V_{GS} = 3$ V and $V_{DS} = 5$ V and compare it to the output conductance at $V_{GS} = 3$ V and $V_{DS} = 0$ V.

Solution. The MOSFET is biased in saturation since $V_{DS} > V_{GS} - V_T$.

Therefore the drain current equals:

$$I_D = \mu_n C_{ox} \frac{W}{L} \frac{(V_{GS} - V_T)^2}{2}$$

$$= 300 \times \frac{3.9 \times 8.85 \times 10^{-14}}{20 \times 10^{-7}} \frac{10}{1} \times \frac{(3-1)^2}{2} = 1.04 \text{ mA}$$

The transconductance equals:

$$g_m = \mu_n C_{ox} \frac{W}{L}(V_{GS} - V_T)$$

$$= 300 \times \frac{3.9 \times 8.85 \times 10^{-14}}{20 \times 10^{-7}} \frac{10}{1} \times (3-1) = 1.04 \text{ mS}$$

and the output conductance equals:

$$g_d = \mu_n C_{ox} \frac{W}{L}(V_{GS} - V_T - V_{DS})$$

$$= 300 \times \frac{3.9 \times 8.85 \times 10^{-14}}{20 \times 10^{-7}} \frac{10}{1} \times (3-1-0) = 1.04 \text{ mS}$$

The measured drain current in saturation is not constant as predicted by the quadratic model. Instead it increases with drain-source voltage due to channel length modulation, drain induced barrier lowering or two-dimensional field distributions, as discussed in section 5.7. A simple empirical model, which considers these effects, is given by:

$$I_{D,sat} = \mu C_{ox} \frac{W}{L} \frac{(V_{GS} - V_T)^2}{2} (1 + \lambda V_{DS}), \text{ for } V_{DS} > V_{GS} - V_T \quad \text{... (5.18)}$$

Where l is a fitting parameter.

5.3.3. The Variable Depletion Layer Model

Next, we develop the variable depletion layer model, which includes the variation of the charge in the depletion layer between the source and drain. This variation is caused by the voltage variation along the channel. The inversion layer charge is still given by:

$$Q_{inv} = -C_{ox} (V_{GS} - V_T), \text{ for } V_{GS} > V_T \quad \text{... (5.19)}$$

where we now include the implicit dependence of the threshold voltage on the charge in the depletion region,

$$V_T = V_{FB} + V_C + 2\phi_F + \frac{\sqrt{2\varepsilon_s q N_a (2\phi_F + V_{SB} + V_C)}}{C_{ox}} \quad \text{... (5.20)}$$

The voltage V_C is the difference between the voltage within the channel and the source voltage. We can now apply the linear model to a small section at a distance y from the source and with a thickness *dy*. The voltage at that point equals $V_C + V_S$ while the voltage across that section equals dV_C. This results in the following expression for the drain current, I_D:

$$I_D = \mu_n C_{ox} \frac{W}{dy} \left(V_{GS} - V_{FB} - 2\phi_F - V_C - \frac{\sqrt{2\varepsilon_s q N_a (2\phi_F + V_{SB} + V_C)}}{C_{ox}} \right) dV_C \quad \text{... (5.21)}$$

Both sides of the equation can be integrated from the source to the drain with *y* varying from 0 to the gate length, L, and the channel voltage, V_C varying from 0 to the drain-source voltage, V_{DS}. This results in:

$$\int_0^L I_D \, dy = \mu_n C_{ox} W \int_0^{V_{DS}} (V_{GS} - V_{FB} - 2\phi_F - V_C) \, dV_C$$

$$- \mu_n W \int_0^{V_{DS}} \sqrt{2\varepsilon_s q N_a (2\phi_F + V_{SB} + V_C)} \, dV_C \quad \text{... (5.22)}$$

Integration yields the following drain current:

$$I_D = \frac{\mu_n C_{ox} W}{L} \left(V_{GS} - V_{FB} - 2\phi_F - \frac{V_{DS}}{2} \right) V_{DS}$$

$$-\frac{2}{3}\mu_n \frac{W}{L}\sqrt{2\varepsilon_s qN_a}\left((2\phi_F + V_{DB})^{3/2} - (2\phi_F + V_{SB})^{3/2}\right) \quad \text{... (5.23)}$$

The current-voltage characteristics as obtained with the above equation are shown in Fig. 5.6, together with those obtained with the quadratic model. Again, it was assumed that the drain current saturates at its maximum value, since a positive inversion layer charge cannot exist in an *n*-type MOSFET. The drain voltage at which saturation occurs is given by:

$$V_{DS,sat} = V_{GS} - V_{FB} - 2\phi_F - \frac{qN_a\varepsilon_s}{C_{ox}^2}\left\{\sqrt{1 + 2\frac{C_{ox}^2}{qN_a\varepsilon_s}(V_{GB} - V_{FB})} - 1\right\} \quad \text{... (5.24)}$$

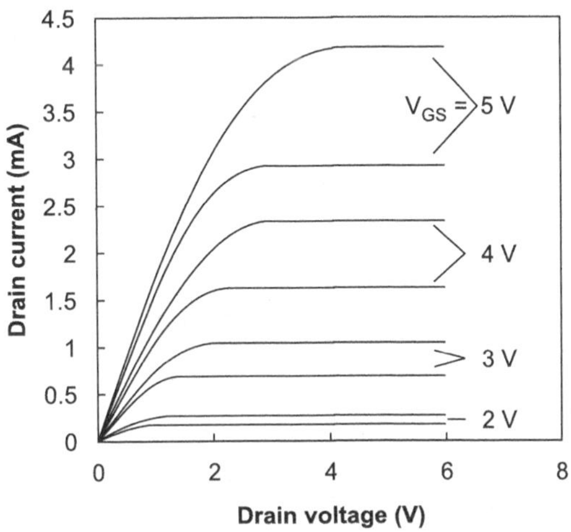

Fig. 5.6 *Comparison of the quadratic model (upper curves) and the variable depletion layer model (lower curves)*

The Fig. 5.6 shows a clear difference between the two models : the quadratic model yields a larger drain current compared to the more accurate variable depletion layer charge model. The transconductance is still given by equation (5.13). This equation combined with the saturation voltage (equation (5.24)) yields:

$$g_{m,sat} = \mu_n C_{ox}\frac{W}{L}\left[V_{GS} - V_{FB} - 2\phi_F - \frac{qN_a\varepsilon_s}{C_{ox}^2}\left\{\sqrt{1 + 2\frac{C_{ox}^2}{qN_a\varepsilon_s}(V_{GB} - V_{FB})} - 1\right\}\right] \quad \text{... (5.25)}$$

This transconductance is almost linearly dependent on V_{GS}, so that it can still be written in the form of equation (5.10) with a modified mobility μ_n^*

$$g_{m,sat} = \mu_n^* C_{ox}\frac{W}{L}(V_{GS} - V_T) \quad \text{... (5.26)}$$

Where μ_n^* equals: $\mu_n^* = \mu_n\left(1 - \frac{1}{\sqrt{1 + \frac{2(2\phi_F + V_{SB})C_{ox}^2}{qN_a\varepsilon_s}}}\right)$ ... (5.27)

The term under the square root depends on the ratio of the oxide capacitance to the depletion layer capacitance at the onset of inversion. Since this ratio is larger than one in most transistors, the modified mobility is 10% to 40% smaller than the actual mobility. This effective mobility can also be used with the quadratic model, yielding a simple but reasonably accurate model for the MOSFET.

Example. 5.2. *Repeat example 5.1 using the variable depletion layer model. Use* V_{FB} = – 0.807 *V and* $N_a = 10^{17} cm^{-3}$.

Solution. To find out whether the MOSFET is biased in saturation, one first calculates the saturation voltage, $V_{D,\,sat}$:

$$V_{DS,\,sat} = V_{GS} - V_{FB} - 2\phi_F - \frac{qN_a\varepsilon_s}{C_{ox}^2}\left\{\sqrt{1 + 2\frac{C_{ox}^2}{qN_a\varepsilon_s}(V_{GB} - V_{FB})} - 1\right\}$$

$$= 1.39 \text{ V}$$

The drain current is then obtained from:

$$I_D = \frac{\mu_n C_{ox} W}{L}\left(V_{GS} - V_{FB} - 2\phi_F - \frac{V_{DS,\,sat}}{2}\right)V_{DS,\,sat}$$

$$-\frac{2}{3}\mu_n\frac{W}{L}\sqrt{2\varepsilon_s qN_a}\left((2\phi_F + V_{DB})^{3/2} - (2\phi_F + V_{SB})^{3/2}\right)$$

$$= 0.7 \text{ mA.}$$

The transconductance equals:

$$g_{m,\,sat} = \mu_n C_{ox}\frac{W}{L}\left[V_{GS} - V_{FB} - 2\phi_F - \frac{qN_a\varepsilon_s}{C_{ox}^2}\left\{\sqrt{1 + 2\frac{C_{ox}^2}{qN_a\varepsilon_s}(V_{GB} - V_{FB})} - 1\right\}\right] = 0.52 \text{ mS}$$

corresponding to a modified mobility $\mu_n^* = 149$ cm^2/V-s. The output conductance at V_{DS} = 0 V equals:

$$g_d = \left.\frac{\partial I_D}{\partial V_{DS}}\right|_{V_{DS}} = 1.04 \text{ mS}$$

Which is the same as that of example 5.1 since the depletion layer width is constant for $V_{DS} = 0$.

5.4 THRESHOLD VOLTAGE

In this section we summarize the calculation of the threshold voltage and discuss the dependence of the threshold voltage on the bias applied to the substrate, called the substrate bias effect.

5.4.1 Threshold Voltage Calculation

The threshold voltage equals the sum of the flatband voltage, twice the bulk potential and the voltage across the oxide due to the depletion layer charge, or:

$$V_T = V_{FB} + 2\phi_F + \frac{\sqrt{2\varepsilon_s q N_a (2\phi_F + V_{SB})}}{C_{ox}} \quad \text{... (5.28)}$$

where the flatband voltage, V_{FB}, is given by:

$$V_{FB} = \Phi_{MS} - \frac{Q_f}{C_{ox}} - \frac{1}{C_{ox}} \int_0^{t_{ox}} \frac{x}{x_{ox}} \rho_{ox}(x)\, dx \quad \text{... (5.29)}$$

With $$\Phi_{MS} = \Phi_M - \Phi_S = \Phi_M - \left(\chi + \frac{E_g}{2q} + \phi_F\right) \quad \text{... (5.30)}$$

and $$\phi_F = V_t \ln \frac{N_a}{n_i}, \text{ } p\text{-substrate} \quad \text{... (5.31)}$$

The threshold voltage of a *p*-type MOSFET with an *n*-type substrate is obtained using the following equations:

$$V_T = V_{FB} - |2\phi_F| - \frac{\sqrt{2\varepsilon_s q N_d (|2\phi_F| - V_{SB})}}{C_{ox}} \quad \text{... (5.32)}$$

where the flatband voltage, V_{FB}, is given by :

$$V_{FB} = \Phi_{MS} - \frac{Q_f}{C_{ox}} - \frac{1}{C_{ox}} \int_0^{t_{ox}} \frac{x}{x_{ox}} \rho_{ox}\, x\,(dx) \quad \text{... (5.33)}$$

With $$\Phi_{MS} = \Phi_M - \Phi_S = \Phi_M - \left(\chi + \frac{E_g}{2q} - |\phi_F|\right) \quad \text{... (5.34)}$$

and $$|\phi_F| = V_t \ln \frac{N_d}{n_i}, \text{ } n\text{-substrate} \quad \text{...(5.35)}$$

The threshold voltage dependence on the doping density is illustrated with Fig. 5.7 for both *n*-type and *p*-type MOSFETs with an aluminium gate metal.

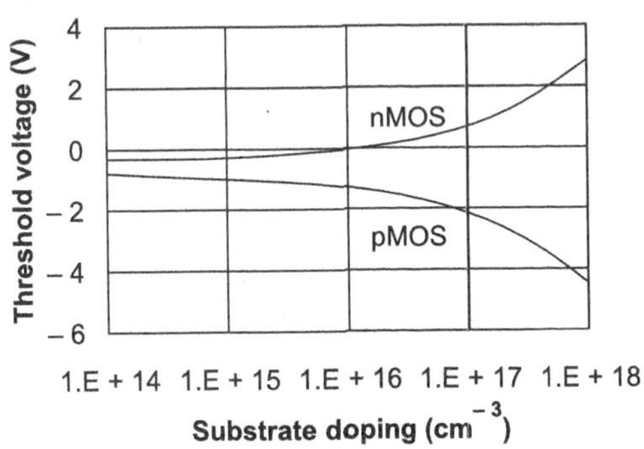

Fig. 5.7 *Threshold voltage of n-type (upper curve) and p-type (lower curve) MOSFETs versus substrate doping density.*

The threshold of both types of devices is slightly negative at low doping densities and differs by 4 times the absolute value of the bulk potential. The threshold of *n*MOSFETs increases with doping while the threshold of pMOSFETs decreases with doping in the same way. A variation of the flatband voltage due to oxide charge will cause a reduction of both threshold voltages if the charge is positive and an increase if the charge is negative.

5.4.2 The Substrate Bias Effect

The voltage applied to the back contact affects the threshold voltage of a MOSFET. The voltage difference between the source and the bulk, V_{BS} changes the width of the depletion layer and therefore also the voltage across the oxide due to the change of the charge in the depletion region. This results in a modified expression for the threshold voltage, as given by:

$$V_T = V_{FB} + 2\phi_F + \frac{\sqrt{2\varepsilon_s q N_a (2\phi_F + V_{SB})}}{C_{ox}} \qquad \text{... (5.36)}$$

The threshold difference due to an applied source-bulk voltage can therefore be expressed by:

$$\Delta V_T = \gamma\left(\left(\sqrt{2\phi_F + V_{SB}}\right) - \sqrt{2\phi_F}\right) \qquad \text{... (5.37)}$$

Where γ is the body effect parameter given by:

$$\gamma = \frac{\sqrt{2\varepsilon_s q N_a}}{C_{ox}} \qquad \text{... (5.38)}$$

The variation of the threshold voltage with the applied bulk-to-source voltage can be observed by plotting the transfer curve for different bulk-to-source voltages. The expected characteristics, as calculated using the quadratic model and the variable depletion layer model, are shown in Fig. 5.8.

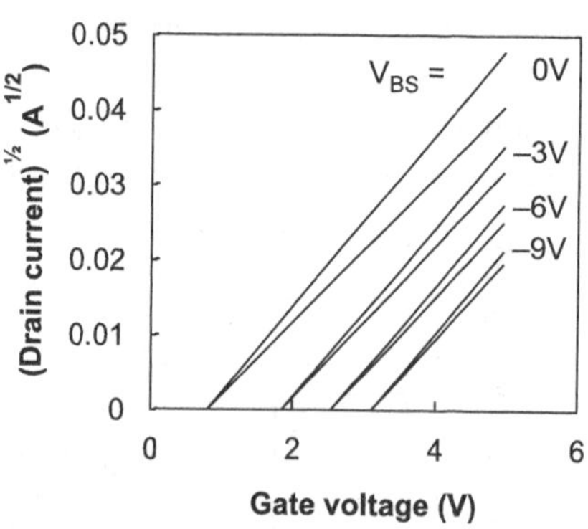

Fig. 5.8 *Square root of I_D versus the gate-source voltage as calculated using the quadratic model (upper curves) and the variable depletion layer model (lower curves).*

First, we observe that the threshold shift is the same for both models. For a device biased at the threshold voltage, drain saturation is obtained at zero drain-to-source voltage so that the depletion layer width is constant along the channel. As the drain-source voltage at saturation is increased, there is an increasing difference between the drain current as calculated with each model. The difference however reduces as a more negative bulk-source voltage is applied. This is due to the larger depletion layer width, which reduces the relative variation of the depletion layer charge along the channel.

Example 5.3. *Calculate the threshold voltage of a silicon nMOSFET when applying a substrate voltage, V_{BS} = 0, – 2.5, – 5, – 7.5 and – 10 V. The capacitor has a substrate doping $N_a = 10^{17}$ cm^{-3}, a 20 nm thick oxide ($\varepsilon_{ox} = 3.9\ \varepsilon_0$) and an aluminium gate (Φ_M = 4.1 V). Assume there is no fixed charge in the oxide or at the oxide-silicon interface.*

Solution. The threshold voltage at V_{BS} = – 2.5 V equals:

$$V_T = V_{T0} + \frac{\gamma}{\sqrt{2\phi_F}}\left(\sqrt{1+\frac{V_{SB}}{2\phi_F}} - 1\right)$$

$$= -\,0.09 + \frac{0.75}{\sqrt{2\times 0.42}}\left(\sqrt{1+\frac{2.5}{2\times 0.42}} - 1\right) = 0.73 \text{ V}$$

Where the flatband voltage without substrate bias, V_{TO}, was already calculated in example 5.2. The body effect parameter was obtained from:

$$\gamma = \frac{\sqrt{2\varepsilon_s q N_a}}{C_{ox}} = \frac{\sqrt{2\times 11.9\times 8.85\times 10^{-14}\times 1.6\times 10^{-19}\times 10^{17}}}{3.9\times 8.85\times 10^{-14}/20\times 10^{-7}}$$

$$= 0.75 \text{ V}^{-1/2}$$

The threshold voltages for the different substrate voltages are listed in the table below.

	V_{BS} = – 2.5 V	– 5 V	– 7.5 V	– 10 V
V_T	0.73 V	1.26 V	1.68 V	2.04 V

5.5 SHORT CHANNEL EFFECT OF MOSFET

A MOS transistor is called a short-channel device if its channel length is on the same order of magnitude as the depletion region thickness of the source and drain junctions.

Alternatively, a MOSFET can be defined as a short channel device if the effective channel length L_{eff} is approximately equal to the source and drain junction depth x_j.

The short channel effects that arise in this case are attributed to four physical phenomena:

(*a*) Velocity saturation

(*b*) Mobility degradation

(*c*) Threshold voltage reduction

(*d*) Drain induced barrier lowering

(*a*) **Velocity Saturation:** The lateral electric field by along the channel increases, as the effective channel length is decreased.

While the electric drift velocity v_d in the channel is proportional to the electric field for lower field values, this drift velocity tends to saturate at high channel electric fields.

The velocity saturation has very significant implications upon the V-I characteristics of the short channel MOS.

The effective channel length L_{eff} will be reduced due to channel-length shortening.

$$I_{D(sat)} = W \cdot v_{d\,sat} \int_0^{L_{eff}} q \cdot n(x)\, dx$$

$$= W\, v_{d\,sat} \mid Q_J \mid \text{ where } Q_j = \int_0^{L_{eff}} q\, n(x)\, dx.$$

Since the channel end voltage is equal to $V_{D\,sat}$, the saturation current can be found as,

$$I_{D(sat)} = W \cdot v_{d\,(sat)}\, C_{ox}\, V_{D\,sat}.$$

(*b*) **Mobility degradation:** In short channel MOS transistors, the carrier velocity in the channel is also a function of the normal (vertical) electric field component E_x.

The surface mobility is reduced with respect to the bulk mobility, due to the scattering of carriers (Collisions suffered by the carriers) in the surface region by influencing the vertical electric field.

The dependence of the surface electron mobility on the vertical electric field can be expressed by the following empirical formula:

$$\mu_{n\,(eff)} = \frac{\mu_{no}}{1 + \Theta E_x} = \frac{\mu_{no}}{1 + \dfrac{\Theta \varepsilon_{ox}}{t_{ox} \cdot t_{si}} \left(V_{os} - V_c(y)\right)}$$

$$\approx \frac{\mu_{no}}{1 + \theta\,(V_{os} - V_T)} \approx \frac{\mu_0}{1 + \theta\,(V_{os} - V_T)}$$

$$\therefore \qquad I_D = \frac{1}{2} \frac{\mu_{no}\, C_{ox}}{1 + \theta\,(V_{os} - V_T)} \cdot \frac{W}{L} \cdot (V_{os} - V_T)^2$$

Assuming,

$\theta\,(V_{os} - V_T) << 1$, we obtain,

$$I_D = \frac{1}{2} \mu_0\, C_{ox} \frac{W}{L} \left[1 - \theta\,(V_{os} - V_T)\right] (V_{os} - V_T)^2$$

$$\approx \frac{1}{2} \mu_0\, C_{ox} \frac{W}{L} \left[(V_{os} - V_T)^2 - \theta\,(V_{os} - V_T)^3\right]$$

where,

μ_{no} = Low-field surface electron mobility = μ_0

Θ = Empirical factor

θ = Empirical coefficient or fitting parameter.

(*c*) **Threshold voltage reduction:** In short-channel MOS transistors, *n*+ drain and source diffusion regions in the *p*-type substrate induce a significant amount of depletion charge.

The threshold voltage is larger for long-channel MOS, than the actual threshold voltage of the short channel MOSFET.

The figure 5.9, is the simplified geometry of the gate-induced bulk depletion region and the *p-n* junction depletion regions in a short-channel MOS transistor.

The bulk depletion region is assumed to have an asymmetric trapizoidal shape, instead of a rectangular shape, to represent accurately the gate induced charge.

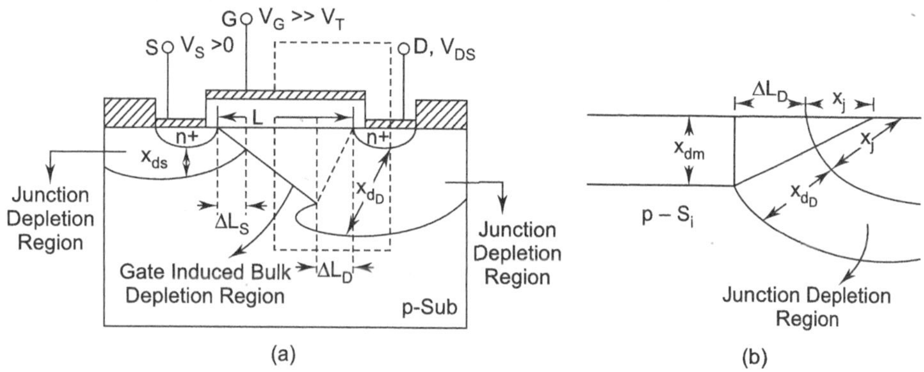

Fig. 5.9 *(a) Simplified geometry of the MOSFET channel region, with gate-induced bulk depletion region and the p-n junction depletion regions. (b) Close-up view of the drain diffusion edge.*

The drain depletion region is expected to be larger than the source depletion region because the *+ve* V_{DS} voltage reverse biases, the drain-substrate junction.

The significant portion of the total depletion region charge under the gate is actually due to the source and drain junction depletion rather than the bulk depletion induced by the gate voltage.

Since the bulk depletion charge in the short channel device is smaller than expected, the threshold voltage expression must be modified to account for this reduction.

Following the modification of the bulk charge term, the threshold voltage of the short channel MOSFET can be written as,

$$V_{TO}\text{ (short channel)} = V_{TO} - \Delta V_{TO}$$

V_{TO} = Zero bias threshold voltage by the conventional formula.

ΔV_{TO} = Threshold voltage reduction.

From the Fig. 5.9(*b*),

The bulk depletion region charge contained within the trapezoidal region is,

$$Q_{BO} = -\left(1 - \frac{\Delta L_S + \Delta L_D}{2L}\right)\sqrt{2q\varepsilon_{si} N_A |2\phi_F|}$$

If, x_{dS} = depth of the *p-n* junction depletion region associated with the source.

x_{dD} = depth of the *p-n* junction depletion region associated with the drain.

The edges of the source and drain diffusion regions are represented by quarter circular arcs, each with a radius equal to the junction depth, x_j.

x_{dm} = the vertical extent of the bulk depletion region into the substrate.

$$x_{ds} = \sqrt{\frac{2}{q}\frac{t_{si}}{N_A}\phi_0} \text{ and } x_{dD} = \sqrt{\frac{2}{q}\frac{t_{si}}{N_A}(\phi_0 + V_{DS})}$$

Junction built in voltage,

$$\phi_0 = \frac{KT}{q}\ln\left(\frac{N_A \cdot N_D}{n_i^2}\right)$$

From Fig. 5.9(*b*),

$$(x_j + x_{dD})^2 = (x_{dm})^2 + (\Delta L_D + x_j)^2$$

or,

$$\Delta L_D^2 + 2x_j \cdot \Delta L_D + x_{dm}^2 - x_{dD}^2 - 2x_j \cdot x_{dD} = 0$$

or,

$$\Delta L_D = -x_j + \sqrt{x_j^2 - \left(x_{dm}^2 - x_{dD}^2\right) + 2x_j \cdot x_{dD}}$$

$$\approx x_j\left(\sqrt{1 + \frac{2x_{dD}}{x_j}} - 1\right)$$

Similarly, we can calculate,

$$\Delta L_s \equiv x_j\left(\sqrt{1 + \frac{2x_{dS}}{x_j}} - 1\right)$$

$$\therefore \quad \Delta V_{TO} = \frac{1}{C_{ox}}\sqrt{2q\,t_{si} \cdot N_A\,|\,2\phi_F\,|} \cdot \frac{x_j}{2L}\left[\left(\sqrt{1 + \frac{2x_{dD}}{x_j}} - 1\right) + \left(\sqrt{1 + \frac{2x_{dS}}{x_j}} - 1\right)\right]$$

(*d*) **Drain Induced Barrier Lowering:** If the channel is very short, then the source and drains depletion layer is overlapped, then the depletion region under the inversion layer increases.

The wider depletion region is accompanied by a larger surface potential which makes the channel more attractive to electrons.

Therefore, a smaller amount of charge on the gate is needed to reach the onset of strong inversion and the threshold voltage decreases. This phenomenon is called Drain Induced Barrier Lowering (DIBL).

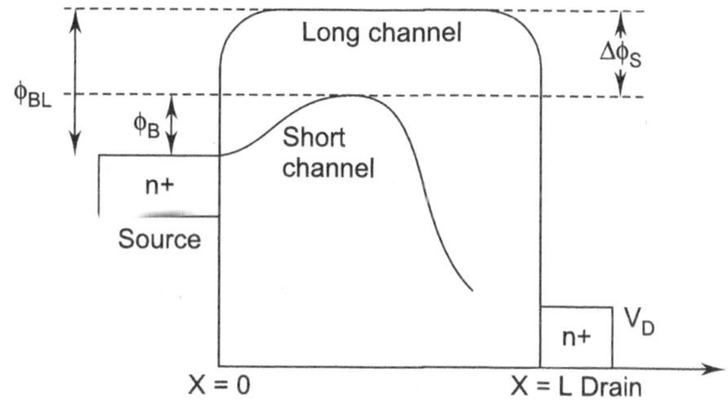

Fig. 5.10

5.6 MOS SCALING

The reduction of the size *i.e.*, the dimensions of the MOSFETs, is commonly referred to as scaling.

Scaling of MOS transistors is concerned with systematic reduction of overall dimensions of the devices as allowed by the available technology, while preserving the geometric ratios found in larger devices.

The proportional scaling of all devices in a circuit would result in a reduction of the total silicon area occupied by the circuit, thereby, increasing the overall functional density of the clip.

All horizontal and vertical dimensions of the large-size transistor are then divided by the scaling factor to obtain the sealed devices. Constant scaling factor, $S > 1$.

There are two types of scaling is possible:

(*a*) Full scaling/constant-field scaling. (*b*) Constant-voltage scaling.

(*a*) **Full-Scaling:** This scaling options aims at preserving the magnitude of the internal electric fields in the MOSFET, while the dimensions are scaled down by a factor of S.

To achieve this goal, all potentials must be scaled down proportionately by the same factor.

Quantity	*Before Scaling*	*After Scaling*
Channel width	W	$W' = W/S$
Channel length	L	$L' = L/S$
Gate oxide thickness	t_{ox}	$t_{ox}' = t_{ox}/S$
Junction depth	x_j	$x_j' = x_j/S$
Power supply voltage	V_{DD}	$V_{DD}' = V_{DD}/S$
Threshold voltage	V_{TO}	$V_{TO}' = V_{TO}/S$
Doping densities	N_A	$N_A' = N_A/S$
	N_D	$N_D' = N_D/S$

Oxide capacitance, $C_{ox}\, e'_{ox} = \frac{t_{ox}}{t_{ox}} = S\left(\frac{\varepsilon_{ox}}{t_{ox}}\right) = SC_{ox}$.

$$I_D'{}_{(lin)} = K_n'\left[\left(V_{ag}' - V_L'\right)V_{DS}' - 0.5\,V_{DS}'^2\right]$$

$$= SK_n \cdot \frac{1}{S^2}\left[\left(V_{ag} - V_T\right)V_{DS} - 0.5\,V_{DS}^2\right] = \frac{I_D\,(lin)}{S}$$

Similarly, $$I_D'{}_{(sat)} = \frac{K_n'}{2} = \left(V_{ag}' - V_T'\right)^2 = \frac{SK_n}{2}\cdot\frac{1}{S^2}\left(V_{ag} - V_T\right)^2$$

$$= \frac{I_D\,(sat)}{S}$$

The instantaneous power dissipated by the device before scaling is, $P = I_D\,.V_{DS}$

After scaling, $P' = I_D'\,.\,V_{DS}' = \frac{V_{DS}}{S}\cdot\frac{I_{DS}}{S} = \frac{P}{S^2}$.

Gate oxide capacitance before scaling, $C_g = WL\,c_{ox}\,X$.

Gate oxide capacitance after scaling, $C_g' = W'\,L'\,C_{ox}'$

$$= \frac{WL\,C_{ox}}{S^3}$$

$$= \frac{C_g}{s^3}.$$

Constant Voltage Scaling

Quantity	Before Scaling	After Scaling
Channel width	W	$W' = W/S$
Channel length	L	$L' = L/S$
Gate-oxide thickness	t_{ox}	$t_{ox}' = t_{ox}/S$
Junction depth	x_j	$x_j' = x_j/S$
Power supply voltage	V_{DD}	remains unchanged
Threshold voltage	V_{TO}	remains unchanged.
Doping demities	N_A, N_D	$N_A' = S^2 N_A$

Linear current, $$I_D'{}_{(lin)} = K_n'\left[\left(V_{ag}' - V_T'\right)V_{DS}' - \frac{1}{2}V_{DS}'^2\right]$$

$$= SK_n\left[\left(V_{ag} - V_T\right)V_{DS} - \frac{1}{2}V_{DS}^2\right] = SI_{D\,(lin)}$$

Saturation current, $$I_D'{}_{(sat)} = \frac{K_n'}{2}\left[\left(V_{ag}' - V_T'\right)^2\right] = S\,I_{D\,(sat)}$$

Power dissipation, $$P' = I_D'\,.\,V_{DS}' = SI_D\,.\,V_{DS} = SP$$

Power density, $$P_D' = \frac{P'}{A'} = \frac{SP}{A/S^2} = S^3\,\frac{P}{A} = S^3\,P_D.$$

Problem 1. *A p-MOS structure is to be made on $N_D = 1 \times 10^{15}$ cm^{-3} Si material. The gate is made of A_l and the gate oxide is 10 nm thick. The Si-SiO_2 interface has a surface charge density of $Q_{SS} = +1 \times 10^{11}$ cm^{-2}.*

(a) *What is the value of ϕ_{ms} for this structure?*

(b) *Calculate the oxide capacitance per unit area.*

(c) *Calculate the flatband voltage for this structure.*

(d) *Calculate the capacitance per unit area at the onset of strong inversion and at high frequency (i.e., when the minority carrier concentration is not significantly affected by the thermal generation/recombination processes).*

Solution. Given datas:

$$n_i = 1.07 \times 10^{10} \text{ cm}^{-3}, \qquad T = 300 \text{ K}$$

$$N_C = 2.84 \times 10^{19} \text{ cm}^{-3} \qquad V_T = 25.852 \text{ mV}$$

$$N_V = 3.08 \times 10^{19} \text{ cm}^{-3} \qquad \phi_{n^+} = \chi_{si} = 4.05 \text{ V}$$

$$q\,\chi_{si} = 4.05 \text{ eV} \qquad \varepsilon_{ox} = 3.9$$

$$Eg_{si} = 1.125 \text{ eV} \qquad \varepsilon_{si} = 11.7$$

(a) Calculation of ϕ_{ms}:

$$\phi_s = \chi_{si} + (E_c - E_v)$$

Now,
$$E_C - E_F = \frac{KT}{q} \ln\left(\frac{N_C}{n}\right) \qquad \text{(where } n = N_D\text{)}$$

$$= 0.026 \times \ln\left(\frac{2.84 \times 10^{19}}{1 \times 10^{15}}\right)$$

$$= 0.265 \text{ V}$$

$\therefore$
$$\phi_{ms} = \phi_m - \phi_s = (4.1 - 4.05 - 0.265) \text{ V}$$

$$= -0.215 \text{ V}$$

(b) Oxide capacitance C_{ox}:

$$C_{ox} = \frac{\varepsilon_{ox}}{t_{ox}} = \frac{3.9 \times 8.85 \times 10^{-14}}{100 \times 10^{-7}} = 3.45 \times 10^{-8} \text{ F/cm}^2.$$

(c) Flatband voltage, V_{fb}:

$$V_{fb}' = \phi_{ms} = -0.215 \text{ V}$$

$\therefore$
$$V_{fb} = V_{fb}' - \frac{Q_{ss}}{C_{ox}} = -0.215 - \frac{1.6 \times 10^{19} \times 1 \times 10^{11}}{3.45 \times 10^{-8}}$$

$$= -0.679 \text{ V}$$

(d) The capacitance/unit area at the onset of strong inversion and at high frequency is C_{min}:

$$\phi_B = \frac{KT}{q} \ln\left(\frac{N_D}{n_i}\right)$$

$$= 0.026 \ln\left(\frac{1\times 10^{15}}{1.07\times 10^{10}}\right) = 0.296$$

$$X_{d_{max}} = \sqrt{\frac{2\varepsilon_{si}\, 2\phi_B}{qN_D}} = \sqrt{\frac{2\times 11.7\times 8.85\times 10^{-14}\times 2\times 0.296}{1.6\times 10^{-19}\times 1\times 10^{15}}}$$

$$= 8.75 \times 10^{-5} \text{ cm.}$$

$$C_{min} = \frac{1}{\frac{1}{C_{ox}} + \frac{X_{d_{max}}}{\varepsilon_i}} = \frac{1}{\frac{1}{3.45\times 10^{-8}} + \frac{8.75\times 10^{-5}}{11.7\times 8.85\times 10^{-14}}}$$

$$= 8.8 \times 10^{-9} \text{ F/cm}^2.$$

$$= 8.8 \ n\text{F/cm}^2.$$

Problem 2. *Doped boron ($N_A = 3 \times 10^{14}$ cm^{-3}) is to be used to fabricate a MOS capacitor. The only contacts are to the gate and the bulk or substrate ($V_B = 0$). The thickness is 350×10^{-7} cm. The gate material is n+ poly si.*

(a) Calculate the gate's work function.
(b) Calculate the p-substrate's work function.
(c) Calculate the Debye length for carrier concentration transition at the oxide-silicon interface.
(d) Calculate the oxide capacitance / cm^2.
(e) Calculate the flatband capacitance / cm^2.
(f) Calculate he flatband voltage.
(g) Calculate the Fermi potential of the substrate material.
(h) Calculate the maximum depletion width at the Si-SiO_2 interface at the onset of strong inversion.
(i) Calculate the depletion charge (cm^{-2}) for the condition in part (h).
(j) Calculate the E-field in Si at the SiO_2 interface.
(k) Calculate the threshold voltage for channel formation in this system.
Where, $\varepsilon_{ox} = 3.9$, $\varepsilon_{si} = 11.7$, T = 300 K, $V_T = 25.852$ mV.

Solution.

(*a*) The gate work-function for *n*+ polysilicon is,

$$\phi_m = \chi_{si} = 4.05 \text{ V}$$

(*b*) Calculation of the *p*-substrate work function:

$$E_F - E_V = \frac{KT}{q} \ln\left(\frac{N_V}{N_A}\right) = 0.026 \ln\left(\frac{3.08\times 10^{19}}{3\times 10^{14}}\right) = 0.298 \text{ V.}$$

Now the work function is,

$$q\,\phi_s = q\,\chi_{si} + Eg - (E_F - E_V)$$

$$= (4.05 + 1.125 - 0.298) \text{ V} = 4.877 \text{ V.}$$

(*c*) Debye length:

$$L_D = \sqrt{\frac{\varepsilon V_T}{qN_A}} = \sqrt{\frac{11.7 \times 8.85 \times 10^{-14} \times 0.026}{1.6 \times 10^{-19} \times 3 \times 10^{14}}} = 2.36 \times 10^{-5} \text{ cm.}$$

(*d*) Calculation of the C_{ox}:

$$C_{ox} = \frac{\varepsilon_{ox}}{t_{ox}} = \frac{3.9 \times 8.85 \times 10^{-14}}{350 \times 10^{-7}} = 9.86\, n\text{F/cm}^2.$$

(*e*) Calculation of the flatband capacitance/cm^2

$$C_{\text{diff}(b)} = \frac{1}{\frac{1}{C_{ox}} + \frac{L_D}{\varepsilon_{si}}} = \frac{1}{\frac{1}{9.86 \times 10^{-9}} + \frac{2.36 \times 10^{-5}}{11.7 \times 8.85 \times 10^{-14}}} = 8.05 \times 10^{-9} \text{ F/cm}^2$$

$$= 8.05\, n\text{F/cm}^2.$$

(*f*) Calculation of the flatband voltage:

$$V'_{fb} = \phi_m - \phi_s = 4.05 - 4.877 = -0.827 \text{ V.}$$

(*g*) Calculation of the Fermi potential of the substrate material:

$$\phi_B = \frac{KT}{q} \ln\left(\frac{n_i}{N_A}\right) = 0.026 \ln\left(\frac{1 \times 10^{10}}{3 \times 10^{14}}\right) = 0.267 \text{ V.}$$

(*h*) Calculation of the maximum depletion width at the Si-SiO_2 interface at the onset of strong inversion:

$$X_{d_{\max}} = \sqrt{\frac{2\varepsilon_{si} 2|\phi_B|}{qN_A}} = \sqrt{\frac{2 \times 11.7 \times 8.85 \times 10^{-14} \times 0.267 \times 2}{1.6 \times 10^{-19} \times 3 \times 10^{14}}}$$

$$= 1.52 \times 10^{-4} \text{ cm.}$$

(*i*) Calculation of the depletion charge (cm^{-2}) for the condition given in part (*h*) :

$$\Phi_{d_{\max}} = -q N_A x_d = -1.6 \times 10^{-19} \times 3 \times 10^{14} \times 1.52 \times 10^{-4}$$

$$= -7.3 \times 10^{-9} \text{ C/cm}^2.$$

(*j*) Calculation of the E-field in the silicon at the SiO_2 interface

$$E_{\max} = -\frac{\Phi_d}{\varepsilon_{si}} = \frac{7.3 \times 10^{-9}}{11.7 \times 8.85 \times 10^{-14}} = 7.07 \text{ KV/cm.}$$

(*k*) Calculation of the threshold voltage for channel formation is this system :-

$$V_T = V_{fb} - 2\phi_B - \frac{\Phi_{d_{\max}}}{C_{ox}} = -0.827 + 2 \times 0.267 + \frac{7.3 \times 10^{-19}}{9.86 \times 10^{-9}} = 0.45 \text{ V.}$$

Problem 3. *Calculate the flatband voltage of a Si n-MOS capacitor with a substrate doping $N_A = 10^{17}$ cm^{-3} and an Al gate ($\phi_m = 4.1V$). Assume there is no fixed charge in the oxide or at the oxide-silicon interface.*

Solution. The flat-band voltage is equal to the work-function difference, since there is no charge in the oxide or at the oxide – semiconductor interface:

$$V_{fb} = \phi_{ms} = \phi_m - \lambda - \frac{E_g}{2q} - V_T \ln\left(\frac{N_A}{n_i}\right) = 4.1 - 4.05 - 0.56 - 0.026 \ln\left(\frac{10^{17}}{10^{10}}\right)$$

$$= -0.93 \text{ V.}$$

Problem 4. *Calculate the threshold voltage of a Si n-MOS capacitor with a substrate doping of $N_A = 10^{17}$ cm^{-3}, a 20-nm-thick oxide ($\varepsilon_{ox} = 3.9\ \varepsilon_0$) and an Al gate ($\phi_m = 7.1$ V). Assume there is no fixed charge in the oxide or at the oxide-silicon interface.*

Solution. Threshold voltage,

$$V_T = V_{fb} + 2\phi_B + \frac{\sqrt{4\varepsilon_s q N_A \phi_B}}{C_{ox}}$$

$$= -0.93 + 2 \times 0.42 + \frac{\sqrt{4 \times 11.9 \times 8.85 \times 10^{-14} \times 1.6 \times 10^{-19} \times 10^{17} \times 0.42}}{3.9 \times 8.85 \times 10^{-14}/20 \times 10^{-7}}$$

$$= -0.09 \text{ V}.$$

Problem 5. *Calculate the drain current of a Si n-MOSFET with $V_T = 1$ V, W = 10 μm, L = 1 μm and $d_{ox} = 20$ nm. The device is biased with $V_{as} = 3$ V and $V_{DS} = 5$ V. Use the quadratic model and a surface mobility of 300 em^2/V-S and set V_{sb} (substrate voltage) = 0 V. Also calculate the transconductance at $V_{as} = 3$ V and $V_{DS} = 5$ V and compare it to the output conductance at $V_{as} = 3$ V and $V_{DS} = 0$V.*

Solution. The MOSFET is biased in saturation since $V_{DS} > V_{as} - V_T$. Therefore, the drain current,

$$I_D = \frac{\mu_e}{2} C_{ox} \frac{W}{L} (V_{as} - V_T)^2 = \frac{300}{2} \times \frac{3.9 \times 8.85 \times 10^{-14}}{20 \times 10^{-7}} \times 10 \times (3-1)^2$$

$$= 1.04 \text{ mA}.$$

The transconductance, $g_m = \mu_e C_{ox} \dfrac{W}{L} (V_{as} - V_T)$

$$= 300 \times \frac{3.9 \times 8.85 \times 10^{-14}}{20 \times 10^{-7}} \times 10 \times (3 - 1 - 0) = 1.04 \text{ ms}$$

The output conductance,

$$g_d = \mu_e C_{ox} \frac{W}{L} (V_{as} - V_T - V_{DS}) = 300 \times \frac{3.9 \times 8.85 \times 10^{-14}}{20 \times 10^{-7}} \times 10 \times (3 - 1 - 0)$$

$$= 1.04 \text{ ms}.$$

Problem 6. *Show that if ρ (x) is the charge density distribution in the oxide region of thickness d_{ox}, the shift in the flatband voltage is given by,*

$$\Delta V_{fb} = -\frac{1}{C_{ox}} \int_0^{d_{ox}} \frac{x\rho(x)\,dx}{d_{ox}}$$

Solution. This problem can be solved by using the principle of superposition. Consider a thin slice of charge with a charge density ρ (x) and spatial width Δ (x) placed a distance x away from the metal in the semiconductor. The electric field produced by this charge is given by Gauss's Law:

$$E(x) = \frac{\rho(x)\Delta(x)}{\varepsilon_{ox}}$$

The charge produced at threshold voltage is,

$$\delta(\Delta V_T) = -\frac{x\rho(x)\Delta(x)}{\varepsilon_{ox}}$$

If the charge density is spread over the entire oxide thickness, the total charge at the threshold voltage, by superposition is,

$$\Delta V_T = \int_0^{d_{ox}} \frac{x\rho(x)\,dx}{\varepsilon_{ox}} = -\frac{1}{C_{ox}}\int_0^{d_{ox}} \frac{x\rho(x)\,dx}{\varepsilon_{ox}}$$

5.7. ADVANCED MOSFET ISSUES

5.7.1 Channel Length Modulation

Channel length modulation in a MOSFET is caused by the increase of the depletion layer width at the drain as the drain voltage is increased. This leads to a shorter channel length and an increased drain current. An example is shown in Fig. 5.11. The channel-length-modulation effect typically increases in small devices with low-doped substrates. An extreme case of channel length modulation is punch through where the channel length reduces to zero. Proper scaling can reduce channel length modulation, namely by increasing the doping density as the gate length is reduces.

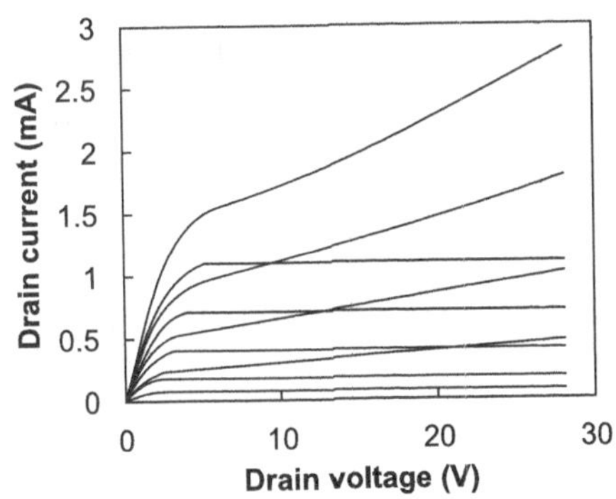

Fig. 5.11 *Current-Voltage characteristics of a MOSFET with and without channel length modulation. ($N_d = 10^{17}$ cm^{-3}, L = 1 μm)*

5.7.2 Drain Induced Barrier Lowering

Drain induced barrier lowering (DIBL) is the effect the drain voltage on the output conductance and measured threshold voltage. This effect occurs in devices where only the gate length is reduced without properly scaling the other dimensions. It is observed as a variation of the measured threshold voltage with reduced gate length. The threshold variation is caused by the increased current with increased drain voltage as the applied drain voltage controls the inversion layer charge at the drain, thereby competing with the gate voltage. This effect is due to the two-dimensional field distribution at the drain end and can typically be eliminated by properly scaling the drain and source depths while increasing the substrate doping density.

5.7.3 Punch Through

Punch through in a MOSFET is an extreme case of channel length modulation where the depletion layers around the drain and source regions merge into a single depletion region. The field underneath the gate then becomes strongly dependent on the drain-source voltage, as is the drain current. Punch through causes a rapidly increasing current with increasing drain-source voltage. This effect is undesirable as it increases the output conductance and limits the maximum operating voltage of the device.

5.7.4 Sub-threshold Current

The basic assumption of the MOS capacitor analysis in section 4.3.2 is that no inversion layer charge exists below the threshold voltage. This leads to zero current below threshold. The actual sub-threshold current is not zero but reduces exponentially below the threshold voltage as:

$$I_D \propto \exp\left(\frac{V_G - V_T}{nV_t}\right) \quad \text{... (5.39)}$$

with

$$n = 1 + \frac{1}{2C_{ox}}\sqrt{\frac{q\varepsilon_s N_a}{\phi_F}} \quad \text{... (5.40)}$$

The sub-threshold behaviour is critical for dynamic circuits since one needs to ensure that no charge leaks through transistors biased below threshold.

5.7.4.1. Derivation of the sub-threshold ideality factor

The charge density below threshold can be expressed as:

$$Q_d = \exp\left(\frac{\phi_s}{V_t}\right) \quad \text{... (5.41)}$$

Where the surface potential, ϕ_s, is related to the gate voltage by:

$$V_G = V_{FB} + \phi_s + V_{ox} = V_{FB} + \phi_s + \frac{\sqrt{2q\varepsilon_s\phi_s}}{C_{ox}} \quad \text{... (5.42)}$$

The gate voltage, V_G, is therefore related to the surface potential, ϕ_s, by :

$$\frac{dV_G}{d\phi_s} = 1 + \frac{1}{2C_{ox}}\sqrt{\frac{2q\varepsilon_s}{\phi_s}} \cong 1 + \frac{1}{2C_{ox}}\sqrt{\frac{q\varepsilon_s}{2\phi_F}} = n \quad \text{... (5.43)}$$

Where the surface potential below threshold was approximated to its value, $2\phi_F$, at threshold. The sub-threshold current therefore equals:

$$I_D \propto Q_d \propto \exp\left(\frac{\phi_s}{V_t}\right) \propto \exp\left(\frac{V_G}{nV_t}\right) \quad \text{... (5.44)}$$

5.7.5 Field Dependent Mobility

The mobility is the inversion layer is distinctly lower than in bulk material. This is due to the fact the electron wavefunction extends into the oxide and the carrier mobility is lowered due to the lower mobility in the oxide. Higher electric fields at the surface-as typically obtained in scaled devices—push the electron wavefunction even more into the oxide yielding a field dependent mobility. The mobility at the surface, $\mu_{surface}$, varies with the electric field, ε, in the following way:

$$\mu_{surface} \propto \varepsilon^{-1/3} \qquad \text{... (5.45)}$$

5.7.6 Velocity Saturation

As devices are reduced in size, the electric field typically also increases and the carriers in the channel have an increased velocity. However at high fields there is no longer a linear relation between the electric field and the velocity as the velocity gradually saturates reaching the saturation velocity. This velocity saturation is caused by the increased scattering rate of highly energetic electrons, primarily due to optical phonon emission. This effect increases the transit time of carriers through the channel. In sub-micron MOSFETs one finds that the average electron velocity is larger than in bulk material so that velocity saturation is not quite as much of a restriction as initially thought.

5.7.7 Oxide Breakdown

As the gate-oxide is scaled down, breakdown of the oxide and oxide reliability becomes more of a concern. Higher fields in the oxide increase the tunneling of carriers from the channel into the oxide. These carriers slowly degrade the quality of the oxide and lead over time to failure of the oxide. This effect is referred to as time dependent destructive breakdown (TDDB).

As the gate-oxide is scaled down, breakdown of the oxide and oxide reliability becomes more of a concern. Higher fields in the oxide increase the tunneling of carriers from the channel into the oxide. These carriers slowly degrade the quality of the oxide and lead over time to failure of the oxide. This effect is referred to as time dependent destructive breakdown (TDDB).

Oxides other than silicon dioxide have been considered as alternate oxides and are typically referred to as high-k dielectrics. These oxides have a larger dielectric constant so that the same gate capacitance can be obtained with a thicker oxide. The challenge is to obtain the same stability, reliability and breakdown voltage as silicon dioxide. Oxides of interest include Al_2O_3, ZrO and TiO.

5.7.8. Scaling

The reduction of the dimensions of a MOSFET has been has dramatic during the last three decades. Starting at a minimum feature length of 10 µm in 1970 the gate length was gradually reduced to 0.15 µm minimum feature size in 2000, resulting in a 13% reduction per year. Proper scaling of MOSFET however requires not only a size reduction of the gate length and width. It also requires a reduction of all other dimensions including the gate/source and gate/drain alignment, the oxide thickness and the depletion layer widths. Scaling of the depletion layer widths also implies scaling of the substrate doping density.

Two types of scaling are common: constant field scaling and constant voltage scaling. *Constant field scaling* yields the largest reduction in the power-delay product of a single transistor. However, it requires a reduction in the power supply voltage as one decreases the minimum feature size. *Constant voltage scaling* does not have this problem and is therefore the preferred scaling method since it provides voltage compatibility with older circuit technologies. The disadvantage of constant voltage scaling is that the electric field increases as the minimum feature length is reduced. This leads to velocity saturation, mobility degradation, increased leakage currents and lower breakdown voltages.

The scaling of MOSFET device parameters is illustrated by Table 5.7.1 where constant field, constant voltage and constant voltage scaling in the presence of velocity saturation are compared.

Table 5.1 Comparison of the effect of scaling on MOSFET device parameters. Compared are constant field scaling, constant voltage scaling and constant voltage scaling in the presence of velocity saturation.

Parameter	*Symbol*	*Constant Field Scaling*	*Constant Voltage Scaling*	*Constant Voltage Scaling with velocity saturation*
Gate length	L	$\frac{1}{\alpha}$	$\frac{1}{\alpha}$	$\frac{1}{\alpha}$
Gate width	W	$\frac{1}{\alpha}$	$\frac{1}{\alpha}$	$\frac{1}{\alpha}$
Field	ε	1	α	α
Oxide thickness	t_{ox}	$\frac{1}{\alpha}$	$\frac{1}{\alpha}$	$\frac{1}{\alpha}$
Substrate doping	N_a	α^2	α^2	α^2
Gate capacitance	C_G	$\frac{1}{\alpha}$	$\frac{1}{\alpha}$	$\frac{1}{\alpha}$
Oxide capacitance	C_{ox}	α	α	α
Transit time	t_r	$\frac{1}{\alpha^2}$	$\frac{1}{\alpha^2}$	$\frac{1}{\alpha}$
Transit frequency	f_T	α	α^2	α
Voltage	V	$\frac{1}{\alpha}$	1	1
Current	I	$\frac{1}{\alpha}$	α	1
Power	P	$\frac{1}{\alpha^2}$	α	1
Power-delay	$P\Delta t$	$\frac{1}{\alpha^3}$	$\frac{1}{\alpha}$	$\frac{1}{\alpha}$

6

Optoelectronic Devices

6.1 INTRODUCTION

Optoelectronics deals with the interaction of electronic processes with light and optical processes. Devices in which such interaction can suitably take place, usually accompanied by an energy conversion process (*e.g.*, from electrical to optical, and *vice versa*), are called optoelectronic devices. Such devices are conveniently made by the semiconductors.

6.2 PHOTODIODES

Photodiodes and crystalline solar cells are essentially the same as the *p-n* diodes, which have been described in this chapter. However, the diode is exposed to light, which yields a photocurrent in addition to the diode current so that the total diode current is given by:

$$I = I_s\left(e^{V_a/V_t} - 1\right) - I_{ph} \qquad \text{... (6.1)}$$

where the additional photocurrent, I_{ph}, is due to photogeneration of electrons and holes shown in Fig. 6.1. These electrons and holes are pulled into the region where they are majority carriers by the electric field in the depletion region.

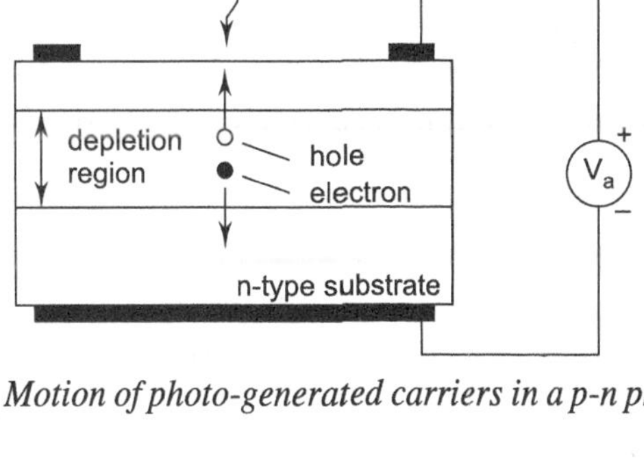

Fig. 6.1 *Motion of photo-generated carriers in a p-n photodiode.*

The photo-generated carriers cause a photocurrent, which opposes the diode current under forward bias. Therefore, the diode can be used as a photodetector—using a reverse or even zero bias voltage—as the measured photocurrent is proportional to the incident light intensity. The diode can also be used as a solar cell—using a forward bias—to generate electrical power.

The primary characteristics of a photodiode are the responsivity, the dark current and the bandwidth. The responsivity is the photocurrent divided by the incident optical power. The maximum photocurrent in a photodiode equals

$$I_{ph,\max} = \frac{q}{h\nu} P_{in} \quad \text{... (6.2)}$$

Where P_{in} is the incident optical power. This maximum photocurrent occurs when each incoming photon creates one electron-hole pair, which contributes to the photocurrent. The maximum photocurrent in the presence of a reflection, R at the surface of the photodiode and an absorption over a thickness d, in a material with an absorption coefficient, α, is given by:

$$I_{ph} = (1 - R)\left(1 - e^{-\alpha d}\right)\frac{qP_{in}}{h\nu} \quad \text{... (6.3)}$$

This photocurrent is obtained by integrating the generation rate over the absorption region with thickness, d. The photocurrent is further reduced if photo-generated electron-hole pairs recombine within the photodiode instead of being swept into the regions where they are majority carriers.

The dark current is the current through the diode in the absence of light. This current is due to the ideal diode current, the generation/recombination of carriers in the depletion region and any surface leakage, which occurs in the diode. The dark current obviously limits the minimum power detected by the photodiode, since a photocurrent much smaller than the dark current would be hard to measure.

However, the true limitation is the shot noise generated by the current through the diode. The shot noise as quantified by the average of the square of the noise current is given by:

$$\left\langle i^2 \right\rangle = 2qI\Delta f \quad \text{... (6.4)}$$

Where I is the diode current and Δf is the bandwidth of the detector. The bandwidth of the diode is affected by the transit time of the photo-generated carriers through the diode and by the capacitance of the diode. The carrier transit time yields the intrinsic bandwidth of the diode while the capacitance together with the impedance of the amplifier or the transmission line connected to the diode yields a the parasitic RC delay.

6.3 SOLAR CELLS

Solar cells are typically illuminated with sunlight and are intended to convert the solar energy into electrical energy. The solar energy is in the form of electromagnetic radiation, more specifically "black-body" radiation. The sun's spectrum is consistent with that of a black body at a temperature of 5800 K. The radiation spectrum has a peak at 0.8 eV. A significant part of the spectrum is in the visible range of the spectrum (400 - 700 nm). The power density is approximately 100 m W/cm^2.

Only part of the solar spectrum actually makes it to the earth's surface. Scattering and absorption in the earth's atmosphere, and the incident angle affect the incident power density. Therefore, the available power density depends on the time of the day, the season and the latitude of a specific location.

Of the solar light, which does reach a solar cell, only photons with energy larger than the energy bandgap of the semiconductor generate electron-hole pairs. In addition, one finds that the voltage across the solar cell at the point where it delivers its maximum power is less than the bandgap energy in electron volt. The overall power-conversion efficiency of single-crystalline solar cells ranges from 10 to 30% yielding 10 to 30 mW/cm^2.

The calculation of the maximum power of a solar cell is illustrated by Figs. 6.1 and 6.2. The sign convention of the current and voltage is shown as well. It considers a current coming out of the cell to be positive as it leads to electrical power generation. The power generated depends on the solar cell itself and the load connected to it. As an example, a resistive load is shown in the diagram below.

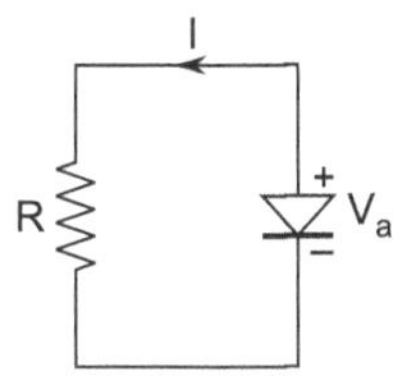

Fig. 6.2 *Circuit diagram and sign convention of a p-n diode solar cell connected to a resistive load. The current and the power as function of the forward bias voltage across the diode are shown in Fig. 6.3 for a photocurrent of 1 mA.*

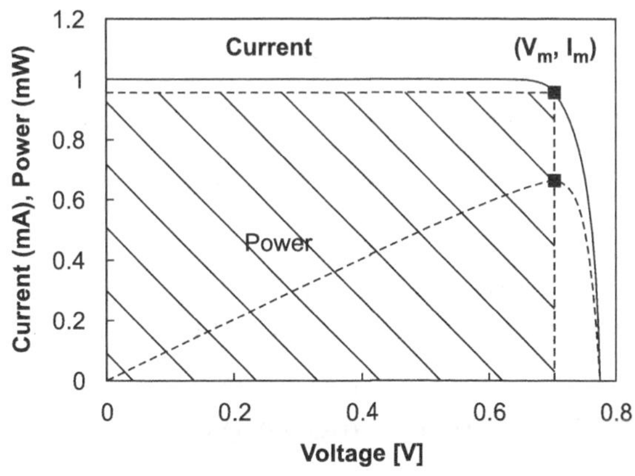

Fig. 6.3 *Current-Voltage (I-V) and Power-Voltage (P-V) characteristics of a p-n diode solar cell with I_{ph} = 1 mA and I_s = 10^{-10} A. The crosshatched area indicates the power generated by the solar cell. The markers indicate the voltage and current, V_m and I_m for which the maximum power, P_m is generated.*

We identify the open-circuit voltage, V_{oc}, as the voltage across the illuminated cell at zero current. The *short-circuit* current, I_{sc}, is the current through the illuminated cell if the voltage across the cell is zero. The *short-circuit* current is close to the photocurrent while the *open-circuit* voltage is close to the turn-on voltage of the diode as measured on a current scale similar to that of the photocurrent.

The power equals the product of the diode voltage and current and at first increases linearly with the diode voltage but then rapidly goes to zero as the voltage approaches the turn-on voltage of the diode. The maximum power is obtained at a voltage labeled as V_m with I_m being the current at that voltage.

The fill factor of the solar cell is defined as the ratio of the maximum power of the cell to the product of the *open-circuit* voltage, V_{oc}, and the *short-circuit* current, I_{sc}, or:

$$\text{Fill Factor} = \frac{I_m V_m}{I_{sc} V_{oc}} \qquad \text{... (6.5)}$$

Example 6.1. *A 1 cm*^{2} *silicon solar cell has a saturation current of* 10^{-12} *A and is illuminated with sunlight yielding a short-circuit photocurrent of 25 mA. Calculate the solar cell efficiency and fill factor.*

Solution. The maximum power is generated for:

$$\frac{dP}{dV_a} = 0 = I_s\left(e^{V_m/V_t} - 1\right) - I_{ph} + \frac{V_m}{V_t} I_s e^{V_m/V_s}$$

where the voltage, V_m, is the voltage corresponding to the maximum power point. This voltage is obtained by solving the following transcendental equation:

$$V_m = V_t \ln \frac{1 + I_{ph}/I_s}{1 + V_m/V_t}$$

Using iteration and a starting value of 0.5 V one obtains the following successive values for V_m:

$$V_m = 0.5,\ 0.542,\ 0.540 \text{ V}$$

and the efficiency equals:

$$\eta = \left|\frac{V_m I_m}{P_{in}}\right| = \frac{0.54 \times 0.024}{0.1} = 13\%$$

The current, I_m, corresponding to the voltage, V_m, was calculated using equation (6.1) and the power of the sun was assumed 100 mW/cm^2. The fill factor equals:

$$\text{Fill factor} = \frac{V_m I_m}{V_{oc} I_{sc}} = \frac{0.54 \times 0.024}{0.62 \times 0.025} = 83\%$$

where the open circuit voltage is calculated using equation (6.1) and I = 0. The short circuit current equals the photocurrent.

6.3.1 The Solar Spectrum

The solar spectrum is shown in Fig. 6.4. The spectrum as seen from a satellite is referred to as the AM0 spectrum (where AM stands for air mass) and closely fits the spectrum of a black body at 5800 K. The total power density is 1353 W/m^2.

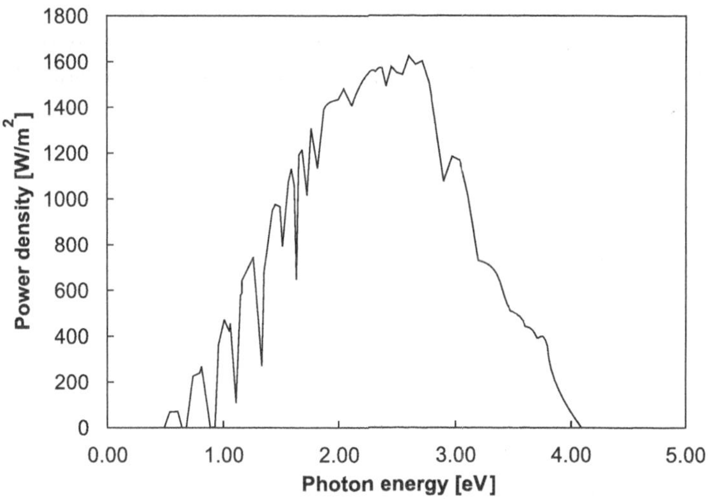

Fig. 6.4 *The solar spectrum under of AM1 conditions.*

The solar spectrum as observed on earth is modified due to absorption in the atmosphere. For AM1 (normal incidence) the power density is reduced to 925 W/cm^2 whereas for AM1.5 (45° above the horizon) the power density is 844 W/m^2. The irregularities in the spectrum are due to absorption at specific photon energies. The corresponding cumulative photocurrent is presented in Fig. 6.5 as a function of the photon energy.

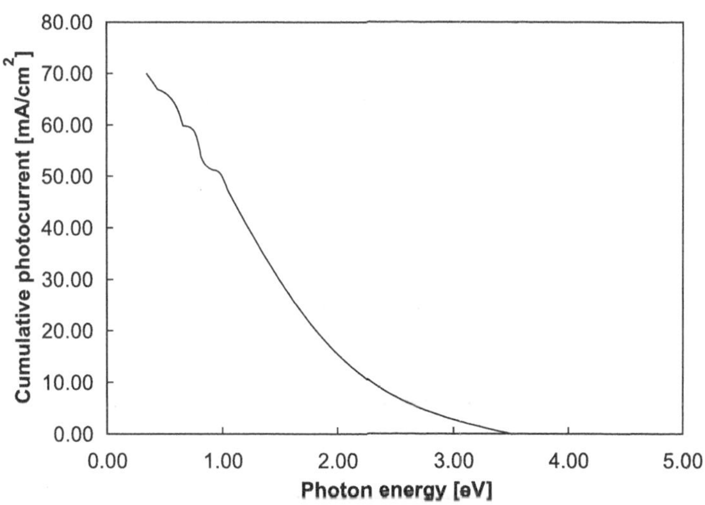

Fig. 6.5 *Cumulative Photocurrent versus Photon Energy under AM1 conditions.*

6.3.2 Calculation of Maximum Power

The current through the solar cell can be obtained from:

$$I = I_s\left(e^{V_a/V_t} - 1\right) - I_{ph} \qquad \text{... (6.6)}$$

where I_s is the saturation current of the diode and I_{ph} is the photocurrent (which is assumed to be independent of the applied voltage V_a). This expression only includes the ideal diode current of the diode, thereby ignoring recombination in the depletion region. The short circuit current, I_{sc}, is the current at zero voltage which equals $I_{sc} = -I_{ph}$. The open circuit voltage equals:

$$V_{oc} = V_a\,(I = 0) = V_t \ln\left(\frac{I_{ph}}{I_s} + 1\right) \cong V_t \ln\frac{I_{ph}}{I_s} \qquad \text{... (6.7)}$$

The total power dissipation is then:

$$P = V_a I = I_s V_a\left(e^{V_a/V_t} - 1\right) - I_{ph} V_a \qquad \text{... (6.8)}$$

The maximum power occurs at $\frac{dP}{dV_a} = 0$. The voltage and current corresponding to the maximal power point are V_m and I_m.

$$\frac{dP}{dV_a} = 0 = I_s\left(e^{V_m/V_t} - 1\right) - I_{ph} + \frac{I_s V_m}{V_t} e^{V_m/V_t} \qquad \text{... (6.9)}$$

This equation can be rewritten as:

$$V_m = V_t \ln\left[\frac{I_{ph} + I_s}{I_s} \frac{1}{1 + \frac{V_m}{V_t}}\right] \cong V_t\left[\frac{V_{oc}}{V_t} - \ln\left(1 + \frac{V_m}{V_t}\right)\right] \qquad \text{... (6.10)}$$

by using equation (6.7) for the open circuit voltage V_{oc}, a more accurate solution is obtained by solving this transcendental equation and substituting into equations (6.9) and (6.9). The maximum power can be approximated by:

$$P_m = I_m V_m \cong -I_{ph}\left(1 - \frac{V_t}{V_m}\right)\left(V_{oc} - V_t \ln\left(1 + \frac{V_m}{V_t}\right)\right) \qquad \text{... (6.11)}$$

$$P_m \cong -I_{ph}\left(V_{oc} - V_t \ln\left(1 + \frac{V_m}{V_t}\right) - \frac{V_{oc} V_t}{V_m}\right) \qquad \text{... (6.12)}$$

or

$$P_m = -I_{ph} \frac{E_m}{q} \qquad \text{... (6.13)}$$

where

$$E_m = q\left(V_{oc} - V_t \ln\left(1 + \frac{V_m}{V_t}\right) - \frac{V_{oc} V_t}{V_m}\right) \qquad \text{... (6.14)}$$

The energy E_m is the energy of one photon, which is converted to electrical energy at the maximum power point. The total photocurrent is calculated as (for a given bandgap E_g)

$$J_{ph}(E_g) = q \quad \text{... (6.15)}$$

and the efficiency equals:

$$\eta = \frac{P_m}{P_n} = \frac{I_{ph}E_m}{qP_{in}} \quad \text{... (6.16)}$$

6.3.3 Conversion Efficiency for Monochromatic Illumination

This first order model provides an analytic approximation for the efficiency of a solar cell under monochromatic illumination. We start with the result of section 6.3.2:

$$\eta = \frac{P_m}{P_{in}} = \frac{I_{ph}}{P_{in}}\left(V_{oc} - V_t \ln\left(1 + \frac{V_m}{V_t}\right) - \frac{V_{oc}V_t}{V_m}\right) \quad \text{... (6.17)}$$

and replace V_{oc} by the largest possible open circuit voltage, E_g/q, yielding:

$$V_m = \frac{E_g}{q} - V_t \ln\left(1 + \frac{E_g}{kT}\right) \quad \text{... (6.18)}$$

and

$$\eta = \frac{I_{ph}V_m}{P_{in}} = \left[1 - \frac{2kT}{E_g}\ln\left(1 + \frac{E_g}{kT}\right)\right] \quad \text{... (6.19)}$$

for a GaAs solar cell at 300K, $\frac{E_g}{kT} = 55$ so that the efficiency equals $\eta = 85\%$

6.3.4 Effect of Diffusion and Recombination in a Solar Cell

6.3.4.1 Photo current versus voltage

$$0 = D_n\frac{d^2n}{dx^2} + \mu_n \varepsilon \frac{dn}{dx} - \frac{n}{\tau} + g_{op} \quad \text{... (6.20)}$$

as well as a similar equation for holes. The photocurrent is obtained from

$$I_{ph}(V_a) = \frac{qA}{d^2}(\phi_i - V_a)\int_0^d (\mu_n n + \mu_p p)\,dx \quad \text{... (6.21)}$$

Once this photocurrent is obtained, the total current is obtained from:

$$I = I_s\left(e^{V_a/V_t} - 1\right) - I_{ph}(V_a) \quad \text{... (6.22)}$$

6.3.5 Spectral Response

Because of the wavelength dependence of the absorption coefficient one expects the shorter wavelengths to be absorbed closer to the surface while the longer wavelengths are absorbed deep in the bulk. Surface recombination will therefore be more important for short wavelengths while recombination in the quasi-neutral region is more important for long wavelengths.

6.3.6 Influence of the Series Resistance

$$V_{ext} = V_a + IR_s \quad \text{... (6.23)}$$

$$I = I_s \left(e^{V_a/V_t} - 1\right) - I_{ph} \quad \text{... (6.24)}$$

$$P = V_{ext} I \quad \text{... (6.25)}$$

Repeating the derivation of section 6.3.2 one can show that the maximum power condition is given by the following set of transcendental equations:

$$V_m = V_t \ln\left[\frac{I_{ph} + I_s}{I_s} \frac{1}{1 + \frac{V_m}{V_t} + \frac{2I_m R_s}{V_t}}\right] \quad \text{... (6.26)}$$

$$I_m = I_s \left(e^{V_m/V_t} - 1\right) - I_{ph} \quad \text{... (6.27)}$$

while the maximum external power equals:

$$P_{m,\,ext} = I_m (V_m + I_m R_s)$$

6.4 LEDS

A light emitting diode consists of a *p-n* diode, which is designed so that radiative recombination dominates. Homojunction *p-n* diodes, heterojunction *p-i-n* diodes where the intrinsic layer has a smaller bandgap (this structure is also referred to as a double-hetero-structure) and *p-n* diodes with a quantum well in the middle are all used for LEDs. We will only consider the *p-n* diode with a quantum well because the analytical analysis is more straightforward and also since this structure is used often in LEDS and even more frequently in laser diodes.

The radiative recombination process is in competition with non-radiative recombination processes such as trap-assisted recombination. Radiative recombination dominates at high minority-carrier densities. Using a quantum well, a thin region with a lower bandgap, positioned at the metallurgical junction, one can obtain high carrier densities at low current densities. These quantum well LEDs have high internal quantum efficiency as almost every electron injected in the quantum well recombines with a hole and yields a photon.

The external quantum efficiency of planar LEDs is much lower than unity due to total internal reflection. As the photons are generated in the semiconductor, which has a high refractive index, only photons traveling normal to the semiconductor-air interface can exit the semiconductor. For GaAs with a refractive index of 3.5, the angle for total internal reflection equals 17° so that only a few percent of the generated photons can escape the semiconductor. This effect can be avoided by having a spherical semiconductor shape, which ensures that most photons travel normal to the interface. The external quantum efficiency can thereby be increased to values larger than 50%.

6.4.1 Rate Equations

The LED rate equations are derived from the continuity equations as applied to the *p-n* diode:

$$\frac{\partial n}{\partial t} = \frac{1}{q}\frac{\partial J_n}{\partial x} - R + G \qquad \text{... (6.28)}$$

where G is the generation rate per unit volume and R is the recombination rate per unit volume. This equation is now simplified by integrating in the direction perpendicular to the plane of the junction. We separate the integral in two parts: one for the quantum well and one for the rest of the structure.

$$\int_{qw} \frac{\partial n}{\partial t} + \int_{p^*c} \frac{\partial n}{\partial t} = \frac{J}{q} - \frac{J_{SHR}}{q} - \frac{J_{bb}}{q} - \frac{J_{new}}{q} - \sum_k \left(N_k P_k - N_k^2\right) B_k - \frac{NP - N_a^2}{N + P + 2N_n \tau_n} \qquad \text{... (6.29)}$$

where *k* refers to the quantum number in the well. IF we ignore the carriers everywhere except in the quantum well and assume that only the first quantum level is populated with electrons/holes and that the density of electrons equals the density of holes, we obtain:

$$\frac{\partial N}{\partial t} = \frac{J}{q} - B_1 N^2 - \frac{N}{2\tau_{nr}} + \frac{S}{\tau_{ab}} \qquad \text{... (6.30)}$$

where the last term is added to include reabsorption of photons. The rate equation for the photon density including loss of photons due to emission (as described with the photon lifetime τ_{ph}) and absorption (as described with the photon absorption time τ_{ab}) equals:

$$\frac{\partial S}{\partial t} = B_1 N^2 - \frac{S}{\tau_{ph}} + \frac{S}{\tau_{ab}} \qquad \text{... (6.31)}$$

The corresponding voltage across the diode equals:

$$V_a = \frac{E_{g,qw}}{q} + V_t \ln\left[\left(e^{N/N_{c,qw}} - 1\right) e^{N/N^*_{v,qw}} - 1\right] \qquad \text{... (6.32)}$$

Where the modified effective hole density of states in the quantum well, $N^*_{v,qw}$, accounts for the occupation of multiple hole levels. The optical output power is given by the number of photons, which leave the semiconductor per unit time, multiplied with the photon energy:

$$P_{out} = h\nu \frac{S}{\tau_{ph}} A(1 - R) \frac{\Theta_c^2}{4} \qquad \text{... (6.33)}$$

where A is the active area of the device, R is the reflectivity at the surface and Θ_c is the critical angle for total internal reflection.

$$R = \left(\frac{n_1 \quad n_2}{n_1 + n_2}\right)^2 \text{ and } \Theta_c = \sin^{-1}\left(\frac{n_1}{n_2}\right) \qquad \text{... (6.34)}$$

The reflectivity and critical angle for a GaAs Air interface are 30% and 16° respectively.

6.4.2 DC Solution to the Rate Equations

The time independent solution (indicated with the subscript 0) in the absence of reabsorption is obtained from:

$$0 = \frac{J_0}{q} - BN_0^2 - \frac{N_0}{2\tau_{ph}} \qquad \text{... (6.35)}$$

$$0 = BN_0^2 - \frac{S_0}{\tau_{ph}} \qquad \text{... (6.36)}$$

where B is the bimolecular recombination constant. Solving these equations yields:

$$N_0 = \frac{1}{4B\tau_{nr}}\left[\sqrt{1+\frac{16\tau_{nr}^2 BJ_0}{q}} - 1\right] \qquad \text{... (6.37)}$$

For small currents this reduces to:

$$(J << q/16\tau_{nr}^2 B)$$

$$N_0 = \frac{2\tau_{nr} J_0}{q} \qquad \text{... (6.38)}$$

which indicates that SHR recombination dominates, whereas for large currents one finds: $(J >> q/16\,\tau_{nr}^2\,B)$

$$N_0 = \sqrt{\frac{J_0}{qB}} \qquad \text{... (6.39)}$$

The dc optical output power is:

$$P_0 = h\nu\, BN_0^2\, A\,(1 - R) \qquad \text{... (6.40)}$$

This expression explains the poor efficiency of an LED. Even if no non-radiative recombination occurs in the active region of the LED, most photons are confined to the semiconductor because of the small critical angle. Typically, only a few percent of the photons generated escape the semiconductor. This problem is most severe for planar surface emitting LEDs. Better efficiencies have been obtained for edge emitting, "super luminescent" LEDs (where stimulated emission provides a larger fraction of photons which can escape the semiconductor) and LEDs with curved surfaces.

6.4.3 AC Solution to the Rate Equations

We now assume that all variables can be written as a sum of a time independent term and a time dependent term (note that $n(t)$ is still a density per unit area):

$$N = N_0 + n_1(t)\; J = J_0 + j_1(t) \qquad \text{... (6.41)}$$

$$S = S_0 + s_1(t)\; P = P_0 + p_1(t) \qquad \text{... (6.42)}$$

$$V_a = V_{a,0} + v_a(t) \qquad \text{... (6.43)}$$

The rate equations for the time-dependent terms is then given by:

$$\frac{\partial n_1}{\partial t} = -B\,2N_0\, n_1(t) - Bn_1^2(t) - \frac{n_1(t)}{2\tau_0} + \frac{j_1(t)}{q} \qquad \text{... (6.44)}$$

$$\frac{\partial s_1}{\partial t} = B\,2N_0\, n_1(t) + Bn_1^2(t) - \frac{s_1(t)}{\tau_{ph}} \qquad \text{... (6.45)}$$

Next, we assuming the ac current of the form $j_1 = j_{1,0}\, e^{j\omega t}$ and ignore the higher order terms. This results in a harmonic solution of the form:

$$n_1 = n_{1,0}\, e^{j\omega t}\; s_1 = s_{1,0}\, e^{j\omega t}\; p_1 = p_{1,0}\, e^{j\omega t} \qquad \text{... (6.46)}$$

with
$$s_{1,0} = \frac{1}{q}\frac{B\,2N_0\, j_{1,0}\, \tau_{ph}\, \tau_{eff}}{\left(1 + j\omega\tau_{ph}\right)\left(1 + j\omega\tau_{eff}\right)} \qquad \text{... (6.47)}$$

where τ_{eff} depends on N_0 as:

$$\tau_{eff} = \frac{1}{2BN_0 + \dfrac{1}{2\tau_0}} \qquad \text{... (6.48)}$$

and the ac responsivity is:

$$\frac{p_{1,0}}{j_{1,0}} = \frac{h\nu}{q}(1 - R)\frac{\pi\Theta_c^2}{4\pi}\frac{B\,2N_0\, \tau_{eff}}{\left(1 + j\omega\tau_{ph}\right)\left(1 + j\omega\tau_{eff}\right)} \qquad \text{... (6.49)}$$

At $\omega = 0$ this also yields the differential quantum efficiency (D.Q.E).

$$\text{D.Q.E} = \frac{p_{1,0}}{j_{1,0}}\frac{q}{hv} = \frac{(1 - R)\,\Theta_c^2\, BN_0\, \tau_{eff}}{2} = \frac{(1 - R)\,\Theta_c^2\, BN_0\, \tau_0}{4BN_0\, \tau_0 + 1} \qquad \text{... (6.50)}$$

6.4.4 Equivalent Circuit of an LED

The equivalent circuit of an LED consists of the *p-n* diode current source parallel to the diode capacitance and in series with a linear series resistance, R. The capacitance, C, is obtained from:

$$\frac{1}{C} = \frac{dV_a}{dQ} = \frac{dV_a}{q\,dN} \qquad \text{... (6.51)}$$

$$\frac{1}{C} = \frac{1}{q}V_t\frac{\left[\dfrac{1}{N_c}e^{N/N_c}\left(e^{N/N_v} - 1\right) + \dfrac{1}{N_v}e^{N/N_v}\left(e^{N/N_c} - 1\right)\right]}{\left(e^{N/N_c} - 1\right)\left(e^{N/N_v} - 1\right)} \qquad \text{... (6.52)}$$

or
$$C = \frac{qN_0}{mV_t} \qquad \text{... (6.53)}$$

with
$$m = \frac{N_0\, e^{N/N_c}}{N_c\left(e^{N/N_c} - 1\right)} + \frac{N_0\, e^{N/N_v}}{N_v\left(e^{N/N_v} - 1\right)} \qquad \text{... (6.54)}$$

$$m = \frac{N_0\left(N_c + N_v\right)}{N_c N_v}$$

for $N_0 << N_c$ and/or N_v, $m = 2$ while for $N_0 >> N_c$ and/or N_v,.

6.5 LASER DIODES

Laser diodes are very similar to LEDs since they also consist of a *p-n* diode with an active region where electrons and holes recombine resulting in light emission. However, a laser diode also contains an optical cavity where stimulated emission takes place. The laser cavity consists of a waveguide terminated on each end by a mirror. As an example, the structure of an edge-emitting laser diode is shown in Fig. 6.6. Photons, which are emitted into the waveguide, can travel back and forth in this waveguide provided they are reflected at the mirrors.

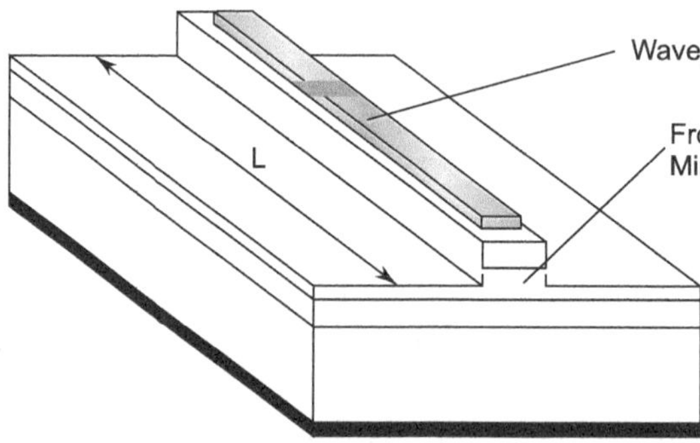

Fig. 6.6 *Structure of an edge-emitting laser diode.*

The light in the waveguide is amplified by stimulated emission. Stimulated emission is a process where a photon triggers the radioactive recombination of an electron and hole thereby creating an additional photon with the same energy and phase as the incident photon. This process is illustrated with Fig. 6.7. This "cloning" of photons results in a coherent beam.

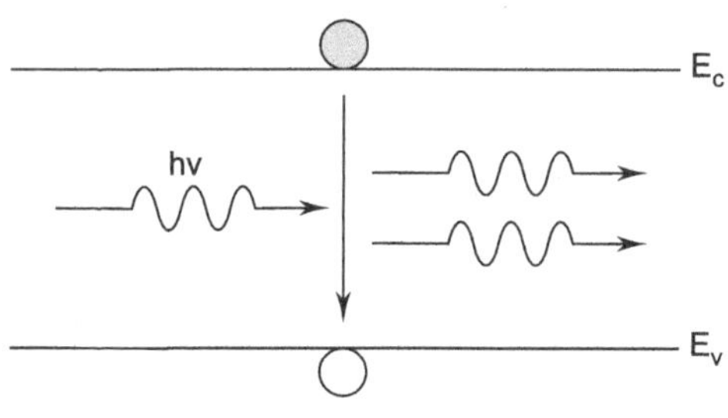

Fig. 6.7 *Stimulated emission of a photon.*

The stimulated emission process yields an increase in photons as they travel along the waveguide. Combined with the waveguide losses, stimulated emission yields a net gain per unit length, g. The number of photons can therefore be maintained if the roundtrip amplification in a cavity of length, L, including the partial reflection at the mirrors with reflectivity R_1 and R_2 equals unity.

This yields the following lasing condition:

$$\text{Roundtrip amplification} = e^{2gL} R_1 R_2 = 1 \quad \text{... (6.55)}$$

If the roundtrip amplification is less than one, then the number of photons steadily decreases. If the roundtrip amplification is larger than one, the numer of photons increases as the photons travel back and forth in the cavity. The gain required for lasing therefore equals:

$$g = \frac{1}{2L} \ln \frac{1}{R_1 R_2} \qquad \text{... (6.56)}$$

Initially, the gain is negative if no current is applied to the laser diode as absorption dominates in the waveguide. As the laser current is increased, the absorption first decreases and the gain increases. The current for which the gain satisfies the lasing condition is the threshold current of the laser, I^{th}. Below the threshold current very little light is emitted by the laser structure. For an applied current larger than the threshold current, the output power, P_{out}, increases linearly with the applied current, as each additional incoming electron-hole pair is converted into an additional photon. The output power therefore equals:

$$P_{out} = \eta \frac{h\nu}{q} (I - I_{th}) \qquad \text{... (6.57)}$$

where hν is the energy per photon. The factor, η, indicates that only a fraction of the generated photons contribute to the output power of the laser as photons are partially lost through the other mirror and throughout the waveguide.

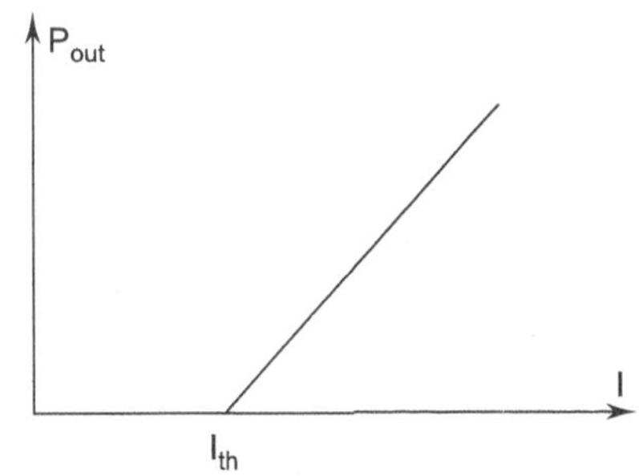

Fig. 6.8 *Output power from a laser diode versus the applied current.*

6.5.1 Laser Cavities and Laser Cavity Modes

A laser diode consists of a cavity, defined as the region between two mirrors with reflectivity R_1 and R_2, and a gain medium, in our case a quantum well. The optical mode originates in spontaneous emission, which is confined to the cavity by the waveguide. This optical mode is amplified by the gain medium and partially reflected by the mirrors. The modal gain depends on the gain of the medium, multiplied with the overlap between the gain medium and the optical mode which we call the confinement factor, Γ, or:

$$\text{modal gain} = g\,(N)\,\Gamma \qquad \text{... (6.58)}$$

Lasing occurs when the roundtrip optical gain equals the losses. For a laser with modal gain g (N) Γ and waveguide loss, α, this condition implies:

$$R_1 R_2 \exp\,[2\,(g\,(N) - \alpha)\,L] = 1 \qquad \text{... (6.59)}$$

where L is the length of the cavity. The distributed loss of the mirrors is therefore:

$$\text{mirror loss} = \frac{1}{L} \ln \frac{1}{\sqrt{R_1 R_2}} \qquad \text{... (6.60)}$$

6.5.1.1 Longitudinal modes in the laser cavity

Longitudinal modes in the laser cavity correspond to standing waves between the mirrors. If we assume total reflection at the mirrors this wave contains $\frac{N}{2}$ periods where N is an integer. For a given wavelength λ and a corresponding effective index, n_{eff}, this yields:

$$N = \frac{2n_{eff} L}{\lambda} \quad \text{... (6.61)}$$

Because of dispersion in the waveguide, a second order modal should also include the wavelength dependence of the effective index. Ignoring such dispersion effects, we find the difference in wavelength between two adjacent longitudinal modes from:

$$N = \frac{2n_{eff} L}{\lambda_1} \quad \text{... (6.62)}$$

$$N + 1 = \frac{2n_{eff} L}{\lambda_2} \quad \text{... (6.63)}$$

$$\Delta\lambda = 2Ln_{eff}\left(\frac{1}{N} - \frac{1}{N+1}\right) \cong \frac{\lambda_1^2}{2Ln_{eff}} \quad \text{... (6.64)}$$

Longer cavities therefore have closer spaced longitudinal modes. An edge emitting (long) cavity with length of 300 μm, n_{eff} = 3.3, and λ = 0.8 μm has a wavelength spacing $\Delta\lambda$ of 0.32 nm while a surface emitting (short) cavity of 3 μm has a wavelength spacing of only 32 nm. These wavelength differences can be converted to energy differences using:

$$\Delta E = \frac{hc}{2Ln_{eff}} = E_{ph} \frac{\Delta\lambda}{\lambda_2} \quad \text{... (6.65)}$$

so that 0.32 nm corresponds to – 6.2 meV and 32 nm to 620 meV. A typical width of the optical gain spectrum is 60 meV, so that an edge emitter biased below threshold can easily contain 10 longitudinal modes, while for a surface emitter the cavity must be carefully designed so that the longitudinal mode overlaps with the gain spectrum.

6.5.1.2 Waveguide Modes

The optical modes in the waveguide determine the effective index needed to calculate the longitudinal modes as well as the confinement factor. Starting from Maxwell's equations in the absence of sources :

$$\vec{\nabla} \times \vec{H} = \varepsilon_0 n^2 (x, y, z) \quad \text{... (6.66)}$$

$$\vec{\nabla} \times \vec{\varepsilon} = \mu_0 \frac{\partial \vec{H}}{\partial t} \quad \text{... (6.67)}$$

and assuming a propagating wave in the z-direction and no variation in the y-direction, we obtain the following one-dimensional reduced wave equation for a time harmonic field, $\varepsilon = \varepsilon_x e^{j\omega t}$, of a TM mode:

$$\frac{d^2 \varepsilon_x}{dx^2} + \left(n^2 (x) k^2 - \beta^2\right) \varepsilon_x = 0 \quad \text{... (6.68)}$$

Where, β, is the propagation constant given by $\beta = \frac{\omega}{c} n_{eff}$, and $k = \frac{\omega}{c}$. The equation then becomes:

$$\frac{d^2 \varepsilon_x}{dx^2} + \left(n^2(x) - n_{eff}^2\right)\varepsilon_x = 0 \qquad \text{... (6.69)}$$

this equation is very similer to the Schrödinger equation. In fact, previous solutions for quantum wells can be used to solve Maxwell's equation by setting the potential V(x) equal to $-n^2(x)$ and replacing $\frac{\omega^2}{c^2}$ by $\frac{2m^*}{\hbar^2}$. The energy eigen values E_l, can then be interpreted as minus the effective indices of the modes: $-n^2_{eff,\,l.}$ One particular waveguide of interest is a slab waveguide consisting of a piece of high refractive index material, n_1, with thickness d, between two infinitely wide cladding layers consisting of lower refractive index material, n_2. From Appendix 17 one finds that only one mode exists for:

$$V_0 = -n_2^2 + n_1^2\ E_{10} = \frac{c^2}{\omega^2}\frac{\pi^2}{d^2} \qquad \text{... (6.70)}$$

or,

$$d \leq \frac{c}{\omega}\pi\frac{1}{\sqrt{n_1^2 - n_2^2}} = \frac{\lambda}{2\sqrt{n_1^2 - n_2^2}} \qquad \text{... (6.71)}$$

For $\lambda = 0.8$ µm, $n_1 = 3.5$ and $n_2 = 3.3$ one finds $d \leq 0.34$ µm.

6.5.1.3 The confinement factor

The confinement factor is defined as the ratio of the modal gain to the gain in the active medium at the wavelength of interest:

$$\Gamma = \frac{\text{modal gain}}{g} = \frac{\int_{-\infty}^{\infty} g(x)\,|\,\varepsilon_x\,|^2\,dx}{\int_{-\infty}^{\infty} |\,\varepsilon_x\,|^2\,dx} \qquad \text{... (6.72)}$$

for a quantum well with width L_x, the confinement factor reduces to

$$\Gamma = \frac{\int_{-L_{x/2}}^{L_{x/2}} |\,\varepsilon_x\,|^2\,dx}{\int_{-\infty}^{\infty} |\,\varepsilon_x\,|^2\,dx} \qquad \text{... (6.73)}$$

which is approximately $\cong 0.02 \ldots 0.04$ for a typical GaAs single quantum well laser.

6.5.2 The Rate Equations for a Laser Diode

Rate equations for each longitudinal mode, λ, with photon density S_λ and carrier density N_λ which couple into this mode are:

$$\frac{\partial N_\lambda}{\partial t} = \frac{J_\lambda}{q} - B_\lambda N_\lambda^2 - \frac{N_\lambda}{2\tau_o} + \sum_k \frac{N_\lambda}{\tau_{k\lambda,k}} - \sum_k^{\lambda} \frac{N_\lambda}{\tau_{k\lambda,k}} - \frac{\partial S_\lambda}{\partial x}\frac{\partial x}{\partial t} \qquad \text{... (6.74)}$$

$$\frac{\partial S_\lambda}{\partial t} = \beta_\lambda B_\lambda N_\lambda^2 - \frac{S_\lambda}{\tau_{ph,\lambda}} + \frac{\partial S_\lambda}{\partial x}\frac{\partial x}{\partial t}, \lambda = 1, 2, ... \lambda_{max} \qquad \text{... (6.75)}$$

Rather than using this set of differential equations for all waveguide modes, we will only consider one mode with photon density S, whose photon energy is closest to the gain peak. The intensity of this mode will grow faster than all others and eventually dominate. This simplification avoids the problem of finding the parameters and coefficients for every single mode. On the other hand it does not enable to calculate the emission spectrum of the laser diode. For a single longitudinal mode the rate equations reduce to:

$$\frac{dN}{dt} = \frac{J}{q} - BN^2 - \frac{N}{2\tau_0} - v_{gr}\Gamma e (N - N_{tr}) S \qquad \text{... (6.76)}$$

$$\frac{dS}{dt} = \beta BN^2 - \frac{S}{\tau_{ph}} + v_{gr}\Gamma e (N - N_{tr}) S \qquad \text{... (6.77)}$$

$$P_1 = v_{gr} SW \ln \frac{1}{\sqrt{R_1}} \qquad \text{... (6.78)}$$

6.5.2.1 DC solution to the rate equations

The time independent rate equations, ignoring spontaneous emission, are:

$$0 = \frac{J_0}{q} - BN_0^2 - \frac{N_0}{2\tau_0} - v_{gr}\Gamma\ell (N_0 - N_{tr}) S_0 \qquad \text{... (6.79)}$$

$$0 = -\frac{S_0}{\tau_{ph}} + v_{gr}\Gamma\ell (N_0 - N_{tr}) S_0 \qquad \text{... (6.80)}$$

where the photon life time is given by:

$$\frac{1}{\tau_{ph}} = \frac{1}{S}\frac{\partial x}{\partial t}\frac{\partial S}{\partial x} = v_{gr}\left(\alpha + \frac{1}{L}\ln\frac{1}{\sqrt{R_1 R_2}}\right) \qquad \text{... (6.81)}$$

from which we can solve the carrier concentration while lasing:

$$N_0 = N_{tr} + \frac{1}{\tau_{ph} v_{gr} \Gamma\ell} \qquad \text{... (6.82)}$$

which is independent of the photon density. The threshold current density is obtained when $S_0 = 0$:

$$J_0 \big|_{(S_0 = 0)} = J_{th} = q\left(BN_0^2 + \frac{N_0}{2\tau_0}\right) \qquad \text{... (6.83)}$$

The photon density above lasing threshold, and power emitted through mirror R_1, are given by :

$$S_0 = \frac{J_0 - J_{th}}{q} \frac{1}{v_{gr}\,\Gamma\,(N_0 - N_{tr})} \qquad \text{... (6.84)}$$

and the power emitted through mirror 1 is :

$$P_{1,0} = h_v S_0 \, W \, v_{gr} \ln \frac{1}{\sqrt{R_1}} \qquad \text{... (6.85)}$$

The differential efficiency of the laser diode is :

$$\text{D.E.} = \frac{dP_{1,0}}{dI_0} = \frac{h\nu}{q} \frac{\ln \frac{1}{\sqrt{R_1}}}{\ln \frac{1}{\sqrt{R_1 - R_2}} + \alpha L} \qquad \text{... (6.86)}$$

and the quantum efficiency is :

$$\eta = \frac{q}{h\nu} \frac{dP_{1,0}}{dI_0} = \frac{\ln \frac{1}{\sqrt{R_1}}}{\ln \frac{1}{\sqrt{R_1 + R_2}} + \alpha L} \qquad \text{... (6.87)}$$

Efficient lasers are therefore obtained by reducing the waveguide losses, increasing the reflectivity of the back mirror, decreasing the reflectivity of the front mirror and decreasing the length of the cavity.

Decreasing the reflectivity of the mirror also increases the threshold current and is therefore less desirable. Decreasing the cavity length at first decreases the threshold current but then rapidly increases the threshold current.

6.5.2.2 AC solution to the rate equations

Assuming a time-harmonic solution and ignoring higher order terms (as we did for the LED) the rate equations become:

$$j\omega n_1 = \frac{j_1}{q} - \frac{n_1}{\tau_{eff}} - v_{gr}\,\Gamma\ell\,(N_0 - N_{tr})\,s_1 - v_{gr}\,\Gamma\ell n_1 S_0 \qquad \text{... (6.88)}$$

$$j\omega s_1 = -\frac{s_1}{\tau_{ph}} + v_{gr}\,\Gamma\ell\,(N_0 - N_{tr})\,s_1 + v_{gr}\,\Gamma\ell n_1 S_0 \qquad \text{... (6.89)}$$

where τ_{eff} is the same as for an LED and given by equation (6.75). Using $v_{gr}\,\Gamma\ell\,(N_0 - N_{tr}) = \frac{1}{\tau_{ph}}$ these equations can be solved yielding:

$$j_1 = j\omega q n_1 + q n_1 \left(\frac{1}{\tau_{eff}} + \Gamma\ell S_0 \, v_{gr} \right) + q n_1 \frac{\Gamma\ell S_0 \, v_{gr}}{j\omega\tau_{ph}} \qquad \text{... (6.90)}$$

We now replace n_1 by relating it to the small signal voltage v_1 :

$$v_1 = \frac{m V_t n_1}{N_0} \qquad \text{... (6.91)}$$

The equation for the small signal current, I_1, can be then written as

$$i_1 = \left(j\omega C + \frac{1}{R} + \frac{1}{j\omega L} v_1 \right) \qquad \text{... (6.92)}$$

with $C = \frac{q N_0 A}{m V_t}$, and $m = \frac{N_0 e^{N/N_c}}{N_c \left(e^{N/N_c} - 1\right)} + \frac{N_0 e^{N/N_s}}{N_v \left(e^{N/N_v} - 1\right)}$,

where A is the area of the laser diode.

$$\frac{1}{R} = C \left(\frac{1}{\tau_{eff}} + \Gamma \ell S_0 v_{gr} \right) \qquad \text{... (6.93)}$$

and

$$L = \frac{1}{C} \frac{\tau_{ph}}{\Gamma \ell S_0 v_{gr}} \qquad \text{... (6.94)}$$

6.5.2.3 Small signal equivalent circuit

Adding parasitic elements and the circuit described by the equation 6.92 we obtain the equivalent circuit of Fig. 6.9, where L_B is a series inductance, primarily due to the bond wire, R_s is the series resistance in the device and C_p is the parallel capacitance due to the laser contact and bonding pad.

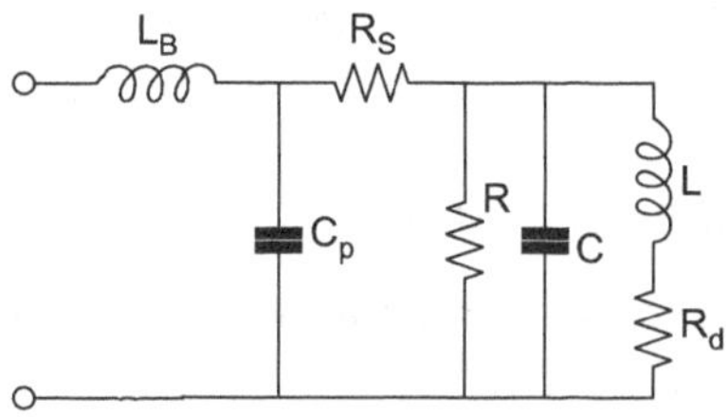

Fig. 6.9 *Small signal equivalent circuit of a laser diode.*

The resistor, R_d, in series with the inductor, L, is due to gain saturation and can be obtained by adding a gain saturation term to equation (6.89). The optical output power is proportional to the current through inductor L, $i_{1,L}$, which is given by:

$$i_{1,L} = \frac{q A s_1}{\tau_{ph}} = q A s_1 v_{gr} \left(\alpha + \frac{1}{L} \ln \frac{1}{\sqrt{R_1 + R_2}} \right) \qquad \text{... (6.95)}$$

and the corresponding power emitted from mirror R_1:

$$p_1 = s_1 h w_{gr} W \ln \frac{1}{\sqrt{R_1}} \qquad \text{... (6.96)}$$

When further ignoring the parasitic elements and the gain saturation resistance, R_d, one finds the ac responsivity $\frac{p_1}{I_1}$ as:

$$\frac{p_1}{i_1} = \frac{h\nu}{q} \frac{\ln \frac{1}{\sqrt{R_1}}}{\ln \frac{1}{\sqrt{R_1 + R_2}} + \alpha L} \frac{1}{1 + j\omega \frac{L}{R} + (j\omega)^2 LC} \quad \text{... (6.97)}$$

from which we find the relaxation frequency of the laser:

$$\omega_0 = \frac{1}{\sqrt{LC}} = \sqrt{\frac{\Gamma \ell S_0 v_{gr}}{\tau_{ph}}} = \sqrt{\frac{\Gamma \ell P_0}{\tau_{ph} h v W \ln \frac{1}{\sqrt{R_1}}}} \quad \text{... (6.98)}$$

The relaxation frequency is therefore proportional to the square root of the DC output power. The amplitude at the relaxation frequency relative to that at zero frequency equals:

$$\frac{p_1 |_{\omega = \omega_0}}{p_1 |_{\omega = 0}} = \frac{R}{L\omega_0} = \frac{1}{\frac{1}{\omega_0 \tau_{eff}} + \tau_{ph} \omega_0} \quad \text{... (6.99)}$$

6.5.3 Large Signal Switching of a Laser Diode

Because of the non-linear terms in the rate equations the large signal switching of a laser diode exhibits some peculiar characteristics. The carrier density initially increases linearly with time while the photon density remains very small since stimulated emission only kicks in for $N > N_0$.

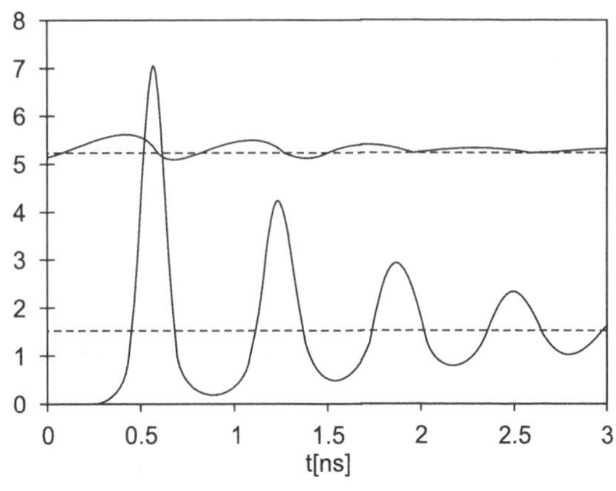

Fig. 6.10 *Optical power and normalized carrier concentration versus time when applying a step current at t =0 from I = 0.95 I_{th} to I = 1.3 I_{th}.*

Both the carrier density and the photon density oscillate around their final value. The oscillation peaks are spaced by roughly $\frac{2\pi}{\omega_0}$, where ω_0 is the small signal relaxation frequency at the final current. The photon and carrier densities are out of phase as carriers are converted into photons due to stimulated emission, while photons are converted back into electron-hole pairs due to absorption. One can use the non-linear behaviour to generate short optical pulses. By applying a current pulse, which is long enough to initiate the first peak in the oscillation, but short enough to avoid the second peak, one obtains a single optical pulse, which is significantly shorter than the applied current pulse. This method is referred to as gain switching or current spiking.

6.6 HETEROJUNCTION BIPOLAR TRANSISTORS

Heterojunction bipolar transistors are bipolar junction transistors, which are composed of at least two different semiconductors. As a result, the energy bandgap as well as all other material properties can be different in the emitter, base and collector. Moreover, a gradual change or "grading" of the material is possible within each region.

Heterojunction bipolar transistors are not just an added complication. On the contrary, the use of heterojunctions provides an additional degree of freedom, which can result in vastly improved devices compared to the homojunction counterparts.

A typical flatband diagram of an HBT and the energy band diagram under forward active bias is presented in Fig. 6.11

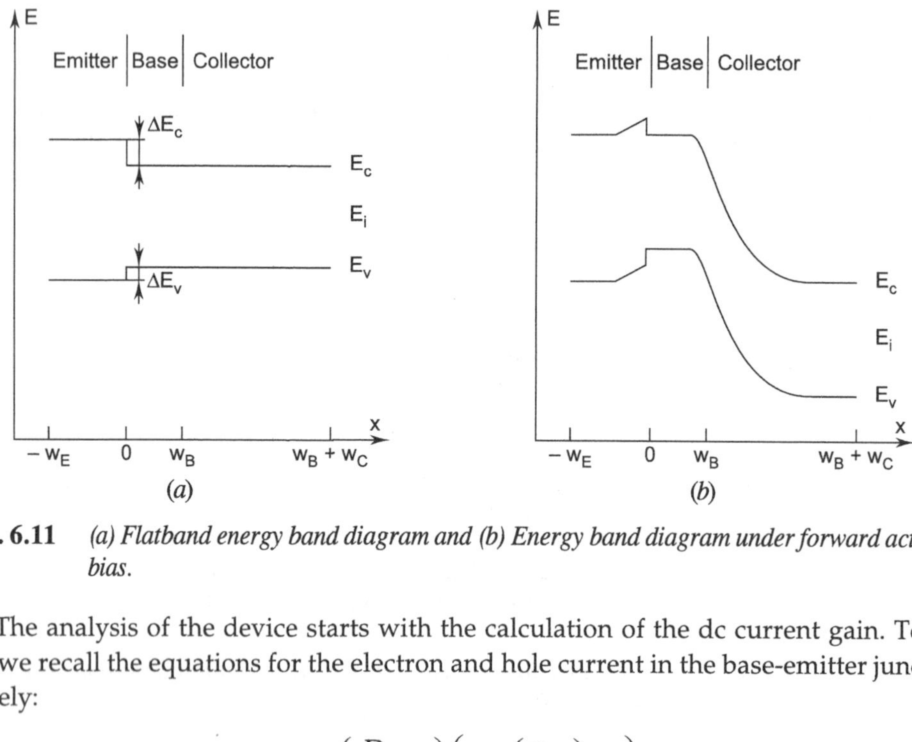

Fig. 6.11 *(a) Flatband energy band diagram and (b) Energy band diagram under forward active bias.*

The analysis of the device starts with the calculation of the dc current gain. To this end we recall the equations for the electron and hole current in the base-emitter junction, namely:

$$I_{E,n} = qn_{i,B}^2 A_E \left(\frac{D_{n,B}}{N_B w_B'}\right)\left(\exp\left(\frac{V_{BE}}{V_t}\right) - 1\right) \quad ...(6.100)$$

and

$$I_{E,p} = qn_{i,E}^2 A_E \left(\frac{D_{p,E}}{N_E w_E'}\right)\left(\exp\left(\frac{V_{BE}}{V_t}\right) - 1\right) \quad ...(6.101)$$

where the fact that the intrinsic carrier density in the emitter, $n_{i,E}$, and base, $n_{i,B}$, are different, is indicated explicitly by the additional subscript.

The emitter147 frequency of the transistor is still calculated from the electron current relative to the total emitter current and equals:

$$\gamma_E \stackrel{\Delta}{=} \frac{I_{E,n}}{I_{E,p} + I_{E,n}} = \frac{1}{1 + \dfrac{D_{p,E} N_B w'_B n_{i,B}^2}{D_{n,B} N_E w'_E n_{i,E}^2}} \qquad \text{... (6.102)}$$

If we now assume that the effective density of states for electrons and holes are the same in the emitter and base, we find that the maximum current gain equals:

$$\beta \cong \frac{D_{n,B} N_E w'_E}{D_{p,E} N_B w'_B} \exp\left(\frac{\Delta E_g}{kT}\right), \text{if} \quad \alpha \cong \gamma_E \qquad \text{... (6.103)}$$

Where ΔE_g is the difference betwen the bandgap energy in the emitter and the bandgap energy in the base. The current gain depends exponentially on this difference in bandgap energy. As a result one can obtain a very large current gain in a heterojunction bipolar transistor, even if the base doping density, N_B, is significantly larger that the emitter doping density, N_E. Therefore, the emitter of a typical heterojunction bipolar transistor has a wide bandgap than its base. This difference should be around 0.2 – 0.4V for optimum performance. A smaller difference provides only a marginal improvement over homojunction devices. A larger difference causes the gain to be strongly temperature dependent and creates a distinct spike in the energy diagram, which in turn limits the current.

The advantages of HBTs are not restricted to its dc performance. Microwave devices can also be improved dramatically by using an appropriate heterojunction material system. To illustrate this point we now recall the equations for the transit frequency, f_T, and the maximum oscillation frequency, f_{max}:

$$f_T = \frac{1}{2\pi\tau} \qquad \text{... (6.104)}$$

Where the total transit time equals:

$$\tau = \tau_E + \tau_B + \tau_C = \frac{V_t C_{BE}}{I_E} + \frac{w_B^2}{2D_n B} + \frac{x_{d,BC}}{2v_{sat}} \qquad \text{... (6.105)}$$

and

$$f_{max} = \sqrt{\frac{f_T}{2\pi R_B C_{BC}}} \qquad \text{... (6.106)}$$

Since a heterojunction transistor can have large gain, even if the base doping density is higher than the emitter doping density, the base can be much thinner even for the same punchthrough voltage. As a result one can reduce the base transit time without increasing the emitter charging time, while maintaining the same emitter current density. The transit frequency can be further improved by using materials with a higher mobility for the base layer and higher saturation velocity for the collector layer.

An HBT can be further improved by grading the composition of the base layer such that the bandgap energy of the material is gradually reduced throughout the base as shown in Fig. 6.12 (*a*) and (*b*). The grading causes an electric field, which in turn reduces the transit time as discussed in section 6.18 and calculated from equation (6.105):

$$\tau_B = \frac{w_B'^{\,2}}{2\left(1+\left(\frac{\varepsilon_{bi}}{\varepsilon_0}\right)^{3/2}\right)D_{n,B}} \quad \text{with } \varepsilon_{bi} = \frac{E_{g,MAX} - E_{g,MIN}}{q w_B'} \qquad \text{... (6.107)}$$

Where $E_{g,\,MAX}$ and $E_{g,\,MIN}$ are the maximum and minimum energy bandgap at $x' = 0$ and $x' = x_B'$. A linear variation was assumed in between.

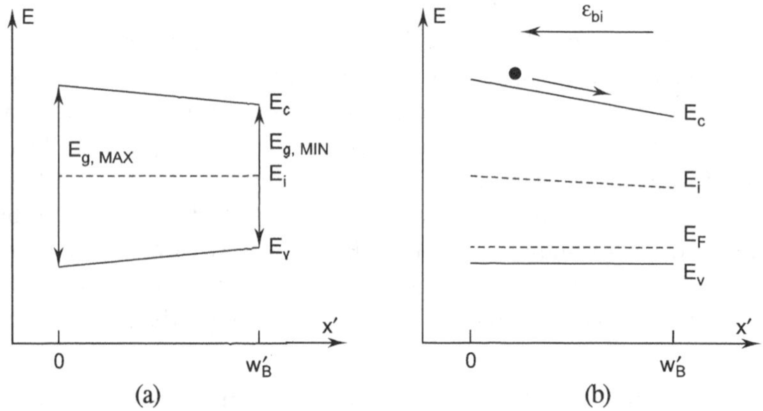

Fig. 6.12 *(a) Energy band diagram of an undoped and composition-graded semiconductor, and (b) Energy band diagram of the p-type graded base region in a HBT*

The maximum oscillation frequency, f_{MAX}, can also be further improved. The improved transit frequency immediately increases f_{MAX}. The higher base doping also provides a lower base resistance and a further improvement f_{MAX}. As in the case of a homojunction BJT, the collector doping can be adjusted to trade off a lower the collector transit time for a lower base-collector capacitance. The fundamental restriction of heterojunction structures still applies, namely that the materials must have a similar lattice constant so that they can be grown without reducing the quality of the material.

7

MEMS

7.1. INTRODUCTION

MEMS means Micro Electro Mechanical Sensors. It is very much applicable in automobiles industries and domestic purposes. The MEMS are mainly depends on stress & strain of the materials by which it is made.

7.1.1 Fundamentals of Mechanics

Consider a solid rod of initial length = L_0, Diameter = D, subjected to a tensile force = F, uniformly applied to the ends of the rod as shown in Fig. 7.1. This tensile force will cause the rod to lengthen by an amount = Δ L.

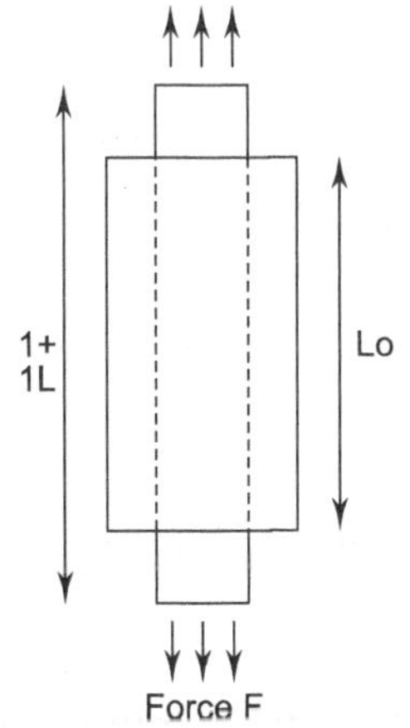

Fig. 7.1 *Rod elongation due to a tensile force.*

Now,

The axial strain $= \varepsilon a$

$$\therefore \qquad \varepsilon a = \left(\frac{\Delta L}{L_0}\right) \qquad \ldots (7.1)$$

Stress $= \sigma$

$$\sigma = \frac{F}{\text{Area}} = \frac{F}{\pi\left(\frac{D}{2}\right)^2} \qquad \ldots (7.2)$$

Strain is a dimensionless quantity, and is usually expressed in units of 10^{-6} or microstrains, because for most materials strain values are quite small.

The units of stress σ are commonly N/m^2 or Pa.

The standard sign convention for stress is for tensile stresses to be negative and compressive stresses to be negative and compressive stresses to be positive.

Here the force F, in figure is acting perpendicularly to the end of the rod. This type of force is termed an axial force, and results in an axial strain εa and axial stress ∂a.

Shear forces, which act parallel to the surface of a body, generate shear stress and strain. From the Hooke's law we get,

$$\partial = \varepsilon E \qquad \ldots (7.3)$$

where E = slope of the stress-strain curve.

Another important mechanical effect is the change in lateral dimensions due to an axial force.

The rod in Fig. 1 is shown as decreasing in diameter as a result of the axial tensile force F. This change in width is characterised by a lateral strain εc, which is related to the axial strain εa by Poisson's ratio v by,

$$v = -\frac{\varepsilon c}{\varepsilon a} \qquad \ldots (7.4)$$

By definition Poisson's ratio is positive. Since εc is a negative value when εa is positive, the negative sign in equation (*iv*) is necessary. For, *si*, $v = 0.28$.

For other material, $v = 0.2$ to 0.5.

7.1.2 Stress in Thin Films

Control of strain levels in deposited thin films is critical for many MEMS devices.

Membrane structures made from polysilicon, a common MEMS mechanical material require tensile film strains less than – 0.001, or the membrane will break.

Tensile stress in a thin film will cause the substrate edges to be higher than the center is viewed from the film side, and a compressive stress will cause the center to be higher.

By measuring the center deflection δ defined as the difference in height between the center and edge, and assuming the bow in uniform across the substrate, the radius of curvature R of the bow can be calculated based upon the size of the substrate (Fig. 7.2). The stress level can then be calculated using the btoney equation :

$$\sigma = \frac{1}{R}\,\frac{E}{\sigma(1-v)}\,\frac{T^2}{t} \tag{7.5}$$

where, R = Radius of curvature

E = Young's modulus

v = Poisson's Ratio

T = Substrate thickness

t = Deposited film thickness.

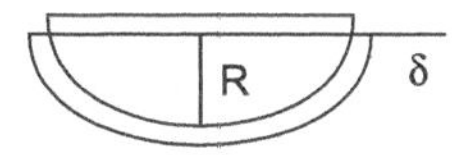

Stress gradients, or variations in stress level as a function of film thickness, are a serious concern in MEMS device fabrication. Stress gradients lead to curvature in cantilever beam and other structures, even if the overall stress level is sufficiently low.

Stress gradients usually result from non-uniformities in the thin film deposition process, which cause atomic structure variations that create uneven strain through the film.

7.1.3 Transduction

MEMS sensors have been fabricated that use many different methods to convert physical changes to an electrical signal.

MEMS sensors generally have a structural element that moves in response to the physical quantity being measured, such as a membrane diaphragm moving in response to a change in applied pressure. The most commonly used method to detect this motion involves piezoresistivity.

The piezoelectric effect is the phenomena whereby a force applied to contain crystalline materials causes an electrical charge to be generated on the surface of the crystal, with the amount of the charge directly related to the applied force. This happens becuase the force creates stress in the crystal, leading to strain.

In piezoelectric materials, the unit cell of the material contains positively and negatively charged ions that are not symmetrically oriented and reflect to each other within the unit cell. The effective location of the center of charge for the positive ions is not at the same location as that of the negative ions. The application of the stress causes deformation (strain), and electric dipoles are created. These electric dipoles induce surface charges on the crystal that in turn create an electric field. This surface charge – induced electric field cancels the stress-induced electric field from the dipoles.

Piezoelectric crystals also show the reverse effect, where an applied electric field causes a mechanical deformation. This effect is important for MEMS actuators.

Piezoelectric materials commonly used in MEMS devices include quartz, Lithium niobate, PZT (lead zirconate titanate), and zinc oxide. These materials can be used both in bulk form (macroscopic disks attached after completion of the MEMS device) or as a thin film deposited during the MEMS fabrication process.

The main advantage of piezoelectric materials is the strong signal transduction for small displacements.

The disadvantage of thin film piezoelectrics is the complication they bring to the device processing. Issues include potential contamination of IC circuitry, nonstandard thin film deposition techniques, and susceptibility to commonly used wet etches.

Capacitance measurements are another common transduction technique for MEMS devices. Consider a pressure sensor made with a flexible membrane that moves in response to changes in the applied pressure. By putting one electrode on the membrane and a second electrode on a fixed surface in close proximity to the membrane electrode, a capacitor is created with capacitance as a function of applied pressure (Fig. 7.2). As the pressure increases/decreases, the membrane moves closer to further from the fixed electrode, and the measured capacitance increases/decreases.

The advantages of capacitance transconduction include limited temp dependence of the measured capacitance and simplicity of the fabrication process.

The major disadvantage is the non-linearity of the dependence of the capacitance on the pressure. This nonlinearity is a result of the membrane deformation mechanics under varying pressure. Accelerometers have also been made with this serving technique.

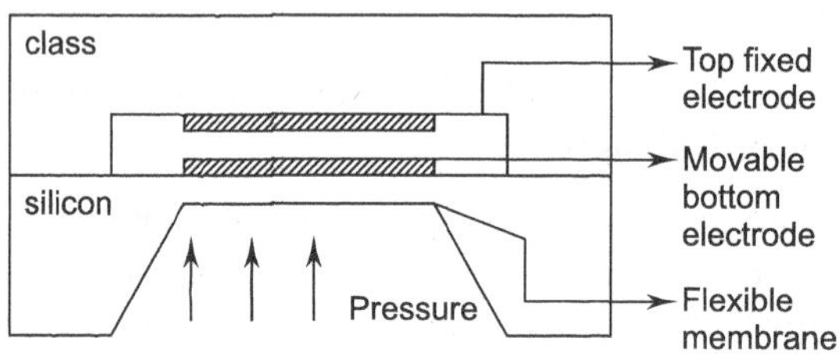

Fig. 7.2

7.1.4 Sensors

Sensors based on the Hall effect generates a voltage proportional to the applied magnetic field. These sensors have been used for detection and position of rotating shafts such as crankshafts in engines. Despite their low Hall Coefficient relative to some other materials, silicon or gallium arsenide are used for much Hall effect MEMS devices because of the developed processing infrastructure for these materials.

7.1.5 Piezoresistivity

Piezoresistivity is a property of materials that describes the change in electrical resistance as a function of mechanical stress applied to the material. Silicon exhibits piezoresistance. The theoretical explanation for piezoresistivity is based on a quantum mechanical description of crystal lattice strain effects on the conductivity of electrons and holes as a function of lattice direction.

From a practical standpoint, piezoresistance in single crystal silicon is primarily dependent on several parameters:

1. The silicon doping type (*n*- or *p*-type) and concentration
2. Temperature.

3. The direction of the current flow relative to the orientation of the crystal lattice.
4. The direction and type of force (tensile or compressive) relative to the orientation of the crystal lattice.

* The most important parameter for MEMS devices made with (100) silicon wafers is the doping type.

Example 7.1 Fig. 7.3 shows a bar of single crystal silicon subjected to a tensile force F, with a current flowing in a direction perpendicular to F. R denotes the resistance of the silicon bar. The force F generates a stress σ in the bar, and a resistance change due to the piezoresistive effect.

The resistance change ΔR is

$$\frac{\Delta R}{R} = \pi_t \sigma_t \qquad \text{... (7.6)}$$

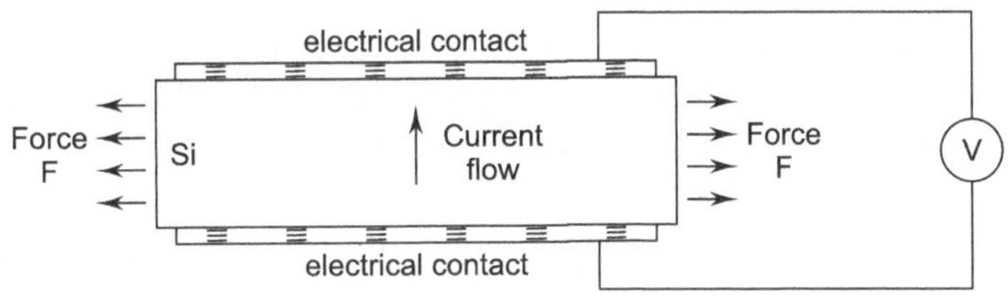

Fig. 7.3 *Current flowing in a si bar with a tensile force applied to the ends of the bar, perpendicular to the current flow*

where

$\pi_t \rightarrow$ Transverse piezoresistance coefficient.

For situations in which the current flow is parallel to the stress, the longitudinal piezoresistance coefficient π_L is used. In the general case in which both transverse and longitudinal stresses are present we have,

$$\frac{\Delta R}{R} = \pi_t \sigma_t + \pi_L \sigma_L \qquad \text{... (7.7)}$$

where σ_t and σ_L are the transverse and longitudinal components of the stress σ, respectively.

For a (100) oriented silicon wafer, planes parallel to the surface have a (100) orientation. Within this plane, the piezoresistive coefficients for silicon with low doping concentration vary as a function of direction.

For *n*-type *si* the piezoresistive coefficients are largest along the $\langle 100 \rangle$ and $\langle 010 \rangle$ directions and smaller along the $\langle 110 \rangle$ directions.

In the case of (100) silicon wafers, these etched features have edges along the $\langle 110 \rangle$ directions. for piezoresistors oriented along the $\langle 110 \rangle$ directions, *p*-type resistors will have higher sensitivity than *n*-type resistors, since piezoresistance values are largest in that direction for *p*-type silicon and smallest for *n*-type silicon.

Inserting the *p*-type silicon piezoresistance coefficients for the $\langle 110 \rangle$ directions into equation (7.7), we find the following for a resistor oriented parallel to the transverse stress:

$$\left(\frac{\Delta R}{R}\right)_{\text{transverse}} = \pi_{p\text{-type}}\,(\sigma_L - \sigma_t) = 70 \times 10^{-11}\ \text{Pa}^{-1}\,(\sigma_l - \sigma_t) \quad \text{... (7.8)}$$

$$\left(\frac{\Delta R}{R}\right)_{\text{longitudinal}} = -\pi_{p\text{-type}}\,(\sigma_l - \sigma_t)$$

$$= -70 \times 10^{-11}\ \text{Pa}^{-1}\,(\sigma_c - \sigma_t) \quad \text{... (7.9)}$$

Each resistor changes by the same amount, but with different sign.

7.1.6 Implementation

The most common implementation of piezoresistors for pressure sensors is to arrange several resistors in a Wheatstone bridge configuration. The Wheatstone bridge effectively cancels the effect of temperature on the piezoresistance if the resistors match each other (have the same resistance value).

Application of positive pressure from above will deflect the membrane downward, induce stress in the membrane, and change the resistance of the four resistors. Due to their orientation, R_1 and R_3 will decrease in resistance, and R_2 and R_4 will increase. Since for p-type resistors,

$$\left(\frac{\Delta R}{R}\right)_{\text{longitudinal}} = -\left(\frac{\Delta R}{R}\right)_{\text{transverse}} \quad \text{... (7.10)}$$

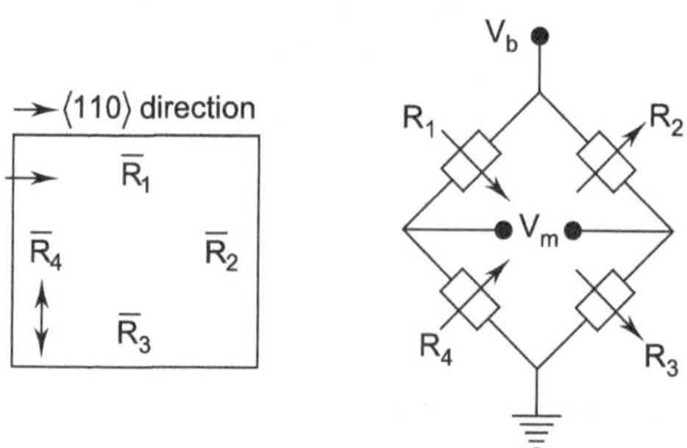

Fig. 7.4

the relative increases in R_2 and R_4 are equal to the relative decreases in R_1 and R_3. If all resistors have the same unstressed resistance value, then a simple circuit analysis of the Wheatstone bridge configuration shows that measurement of the two voltages V_b and V_m determines $\frac{\Delta R}{R}$ by the relation,

$$\frac{\Delta R}{R} = \frac{V_b}{V_m} \quad \text{... (7.11)}$$

The two major advantages of piezoresistive MEMS device :

(i) The simplicity of processing and the maturity of measurement process.
Silicon is by its nature piezoresistive, and by adding dopants, resistors can easily be created either using diffusion or ion implantation.

(ii) Measurement of resistance changes through use of Wheatstone bridge configurations to convert the resistance change to a voltage is well established.

Disadvantage

The piezoresistive sensors is the $\frac{1}{T}$ temperature dependence of the piezoresistance coefficients for single crystal silicon. This signifies temperature dependence can require temperature compensation circuitry, which complicates the implementation of piezoresistive based MEMS device.

7.2 MECHANICS OF COMMON MEMS DEVICES

Many MEMS devices utilize either thin film membrane or cantilever beam structures. The mechanism of thin film membranes is based on the mechanics of thin plates (sometimes called shells), whereas cantilever beam structures are based upon the mechanics of beams.

Thin film membranes in MEMS devices are usually squares or rectangular in shape, with side lengths ranging from hundreds of microns to more than several millimeter less common are circular membranes, primarily due to the difficulty of etching circular structures. The membrane material is usually silicon (either single crystal or polysilicon) or silicon nitride. Membrane thicknesses depend on the area of the membrane and the application, and can range from a few thousand angstroms to 50 μm or more. MEMS cantilever beams are commonly hundreds of microns long, tens to hundreds of microns wide, and less than five microns in thickness. Note that for both membranes and cantilever beam MEMS structures, the thickness is much less than the lateral dimensions of the structure.

The important mechanical properties of membranes and cantilever beams can be derived from a stress/strain analysis using thin plate deflection theory or beam deflection theory. Although a complete discussion of these analysis is beyond the scope of this book, the significant results for MEMS devices can be summarized. For membranes, the relevant assumptions of thin plate deflection theory are that (1) the maximum membrane displacement is less than 20% of the membrane thickness, (2) membrane thickness does not exceed 10% of plate length, and (3) there is no initial stress in the membrane. The first assumption puts a limit on the force that can be applied to the membrane. The second assumption is easily met for most MEMS structures, where the membrane thickness is generally less than 50 μm, which means the membrane size must be at least 0.5 mm on a side. If either of the first two assumptions is not met, thick plate theory must be used. If the third assumption is not met, then the results are qualitatively correct, but a more complete analysis is required to obtain quantitatively accurate values. Although real

MEMS films usually have some intrinsic stress, for simplicity we assume no intrinsic stress.

For a square membrane of side a, thickness t, Young's Modulus E, density ρ and Poisson's ratio v, subjected to a uniform pressure P. The maximum membrane deflection W_{max}, maximum longitudinal and transverse stress σ_l and σ_t, and fundamental mode resonant frequency of vibration F are defined as :

Max deflection $W_{max} = 0.001265\ Pa^4/D$

Max longitudinal stress $\sigma_l = 0.3081\ P\,(a/t)^2$

Max transverse stress $\sigma_t = v\,\sigma_l$

Resonant frequency $$F_o = \frac{1.654t}{a^2}\left[\frac{E}{\rho\left(1-v^2\right)}\right]^{\frac{1}{2}}$$

where the plate fractual rigidity D, a measure of stiffness of the plate to binding, is defined as

$$D = \frac{Et^3}{12\left(1-v^2\right)}$$

The maximum longitudinal stress σ_l is at the centre of each side at the edge of the membrane, and perpendicular to the membrane edge. The maximum transverse stress σ_t is also at the center of the side along the edge, but is oriented parallel to the edge. The location of the maximum stress points is important for piezoresistive and piezoelectric devices, since maximum stress leads to maximum effect when using these transduction methods.

For cantilever beam theory, the major assumption is that the deflection at the end of the beam is small compared with the beam length. For a cantilever fixed at one end, length L, width a, thickness t, Young's modulus E, density ρ, and uniform distributed load P (P is a force per length) as shown in Fig. 7.5.

Deflection $$W(P_1\,x) = \frac{Px^2}{24EI}\left(6L^2 - 4Lx + x^2\right)$$

Max stress $$\sigma = \frac{PL^2t}{4I}$$

where I is the binding moment of inertia defined as $I = \dfrac{at^3}{12}$.

For a uniform force F applied to the beam surface, the load P is equal to the force F scaled by the beam width a, so

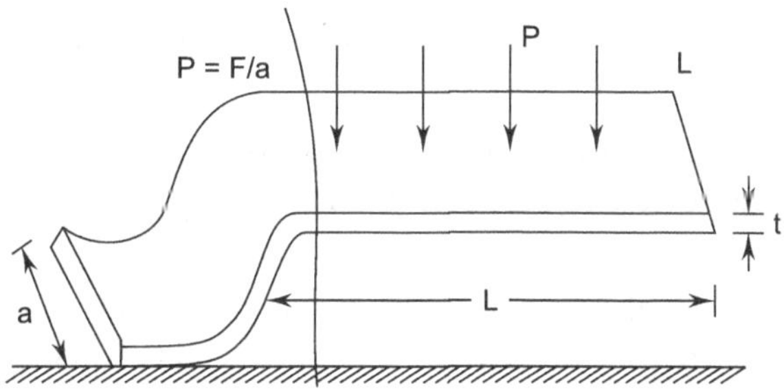

Fig. 7.5 *Cantilever beam fixed at one end, subjected to downward force per unit length P.*

For a point load Q (Q is a force) at the end of the beam

$$\text{Deflection W (P, } x) = \frac{Qx^2}{6EI}(3L - x)$$

$$\text{Max stress} \qquad \sigma = \frac{QLt}{2I}$$

The fundamental mode resonant frequency of the beam is

$$F_0 = 0.161 \frac{t}{L^2}\left(\frac{E}{\rho}\right)^{\frac{1}{2}}.$$

Note that F_0 does not depend on the beam width.

In terms of the cantilever beam mass M.

$$F_0 = 0.161 \frac{t}{L}\left(\frac{Eta}{ML}\right)^{\frac{1}{2}}.$$

For the cantilever beam, the maximum stress is at the base of the beam and is longitudinal. The transverse stress for the beam is zero because unlike the membrane where all sides are fixed, the sides of the beam are free to move up and down, so there is no force transverse to the beam for simple up and down motion. More complex cantilever motions involving torsional motions require more complicated analysis.

7.3 BULK MICROMACHINING ETCHING TECHNIQUES

Bulk micromachining refers to MEMS fabrication processes that involve removal of significant amounts of the silicon substrate in order to form the desired structure. Etching is the cornerstone of bulk micromachining.

Wet etching with both isotropic and anisotropic etchants has dominated MEMS devices, but more recently other techniques, such as isotropic vapor phase etching and high density plasma-based processes, have been used.

Isotropic Etching : Isotropic wet etchants for silicons such as mixtures of hydrofluoric acid-nitric acid-acetic acid, show each profiles that are difficult to control. The under cutting of features defined by the masking material leads to rounded profiles for holes and trenches. Fig. 7.6 shows each profiles for isotropic etchants with and without agitation of the etchant.

With agitation, the profile generally exhibits the shape expected to etch rates that are nearly the same for all crystal orientations.

Without agitation, the profile shows a flatter bottom, which is the result of reduced etching in the vertical direction. This reduction in etch rate is caused by a depletion of the etching species near the etch surface (the etch is diffusion-rate limited). The agitation assists the diffusion of fresh etchant to the surface, resulting in the expected isotropic etch profile.

For bulk micromachined structures, etch depths often approach the full wafer thickness. When isotropically etching there deep structures, the considerable undercutting of the etch mark means the features must be separated by at least the depth of the etch.

Factors to limiting the use of isotropic etchants in MEMS processing :-

(*i*) Sensitivity of the etch profile to agitation level.
(*ii*) Significant amount of undercutting for marked features.

XeF_2 : XeF_2 vapour has been shown to etch silicon without any excitation such as heat or plasma.

Advantage : High etch selectivity to Al, SiO_2, Si_3N_2 and photoresist and a simple etch system.

Disadvantage : Very rough silicon surface resulting from the etch.

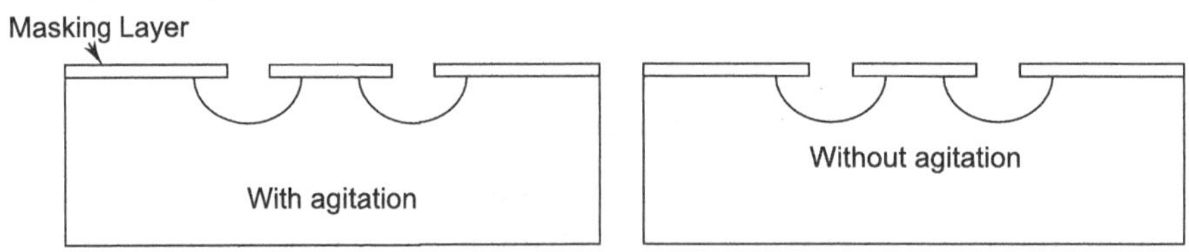

Fig. 7.6

Isotropic etch cross sections showing undercutting of the etch mark and the effect of agitation on the etch profile.

Anisotropic Etching : Anisotropic etching permits the formation of features defined by the crystal planes in the silicon wafer.

KOH : It is the common anisotropic etchant for bulk silicon micromachining.

EDP : Ethylene diamine pyrocatechol (EDP) mixed with water, hydrazine (N_2H_4) and water and tetramethyl ammonium hydroxide (TMAH).

These etches are called anisotropic because the etching rate in the $\langle 100 \rangle$ direction is high and in the $\langle 111 \rangle$ direction is low.

Because (1) the fact that the $\langle 111 \rangle$ crystal plane is the most dense, with three of the four covalent bands below the plane, and thus less accessible to the etchant, and (2) surfaces with the highest bond density etch fastest.

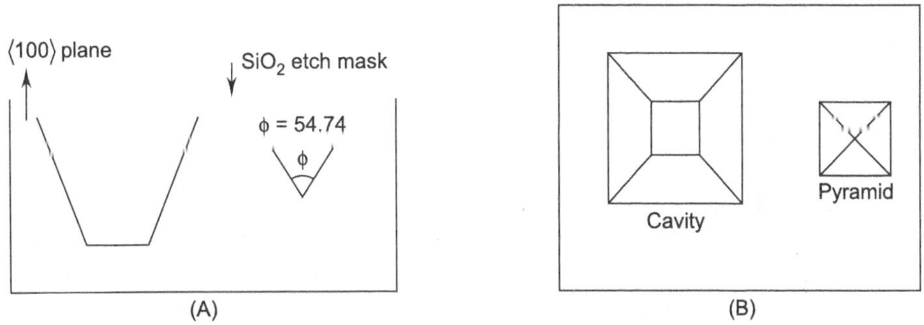

Fig. 7.7 *Cross Section (A) and top view (B) of pryamidal holes and cavities formed in a ⟨ 100 ⟩ silicon wafer with an anisotropic etchant*

KOH : It is commonly used at temperatures above 60°C and at concentrations of 20 to 50% by weight in water.

Advantage : It is relatively safe and simple.

Disadvantages :

(*i*) High SiO_2 etching rate. This means thick SiO_2 films are needed to act as a mark for deep etches.

(*ii*) Due to potassium, is not compatible with conventional IC processing unless the circuit area can be adequately protected.

EDP, $N_2 H_4$ / Water, TMAH :-

Advantages :

(*i*) Low SiO_2 etch rate.

(*ii*) Most metals are not etched by $N_2 H_4$.

(*iii*) TMAH is relatively safe and easy to handle with excellent selectivity to SiO_2 and Si_3N_4.

Disadvantages :

(*i*) N_2H_4 / water is potentially explossive.

(*ii*) N_2H_4 is also a suspected careinogen.

(*iii*) The etch ratio between the ⟨ 100 ⟩ and ⟨ 111 ⟩ planes is not as high as for the other anisotropic etchants.

7.4 ASPECT OF MEMS BULK MICROMACHINING

An important aspect of MEMS bulk micromachining is the capability of forming structures with reproducible mechanical properties such as resonant frequency and deflection.

Since the lateral dimensions are generally determined by photolithographic tolerances, which are quite accurate relative to the size of MEMS devices, thickness is the dimension most difficult to control.

Most bulk micromachined structures have their thickness determined by wet etching, so control of etch depth is the key to device yield and performance methods for controlling etch depth in bulk micromachining are called etch stop techniques.

7.4.1 Four Common Techniques Exist

1. Timed etches.
2. Anisotropic etching of V grooves
3. p^{++} doping
4. Electromechanical etch stop.

1. **Timed Etches :** Timing the etch is the simplest, though least accurate technique. By knowing the etch rate, and the desired etch depth, the etching time can be calculated. One problem with timed etches is the loading effect, where the etch rate changes as silicon is etched, due to dilution of the etch species. This leads to slower etch rates as a function of time. Different feature sizes often etch at different rates due to diffusion effects. Also, variations in initial wafer thickness across the wafer are directly translated to thickness differences of the etched features.
2. **Anisotropic Etching of V Grooves :** It is another etch stop technique that is simple from a processing standpoint. Using simple geometry, the depth of a rectangular pattern can be calculated for a KOH etch. The dimensions of the rectangle are chosen such that the flat bottom of the trench disappears at the desired etch depth, yielding a V-groove. While etching, close monitoring of the rectangular area indicates when the etch is complete. Disadvantages of this technique include the necessity of stopping the etch at the moment indicated by the rectangular V-groove, and the need for wafers of low TTV.
3. **p^{++} doping etching :** (The p^{++} doping etch stop technique is based on the fact that all anisotropic etchants show a drastic decrease in etch rate when the silicon is heavily doped with boron) Etch rates start to drop at boron concentration in the range of $10^{19}/\text{cm}^3$, with the etch rate dropping by a factor ranging from 20 to 50 dpending on the etchant. (One proposed mechanism for this effect is that the heavy boron doping reduces the thickness of the space charge region near the surface, and electrons can more easily tunnel through this region and recombine with holes within the silicon. This limits the availability of the electrons necessary for the etching reaction and slows the rate.

   Heavily doped boron regions with thicknesses up to 20 μm can be formed by diffusion and thinner regions with ion implantation. The high boron doping creates a tensile stress in the film resulting from the smaller size of the boron atom.
4. **Electromechanical Etch Stop :** Electromechanical etch stop techniques are based on the fact that electrically biasing *n*-type silicon positively with respect to the etchant solution prevents the n-type silicon from etching. The electric field induced in the *n*-region interferes with the chemical reaction involving the hydroxide ions that oxidize the silicon surface. A common usage of this effect is to create an *n*-type region on a *p*-type substrate using epitaxial growth or ion implantation. The thickness of the *n* region corresponds to the desired membrane or cantilever beam thickness. Using an electrochemical etching setup. (Fig. 7.8), proper biasing of the Si will cause the anisotropic KOH etch to stop when the *p*-type *Si* is gone from the etch hole.

For a piezoresistive pressure sensor with low doped *p*-type piezoresistive elements on the surface of the *n*-type region, this technique is preferable to the p^{++} doping etch stop since the piezoresistance coefficients for the low doped *p* region are higher than for a p^{+} region. Also, the silicon membrane quality is higher for the electrochemically etched structure since there is no membrane strain induced by the high p^{++} doping.

Advantages :

1. lack of crystal orientation effect.
2. vertical features leading to closer spacing of structures.
3. high aspect ratios, allowing the formation of structures impossible using wet etching

Disadvantages of p^{++} doping techniques

1. The piezoresistance coefficients decrease significantly at these doping levels.
2. The high boron concentration is not compatible with IC technologies, so integrations with on chip electronics is difficult.

Disadvantages of Electrochemical Etching

1. Capital cost of the equipment is large.
2. Production systems are considerably more expensive.

7.5 BULK MICROMACHINING PROCESS FLOW OF A PIEZORESISTIVE PRESSURE SENSOR

The piezoresistive pressure sensor may be used to measure air pressure in an automobile air intake system. Fig. 7.8 shows a schematic drawing of such a sensor.

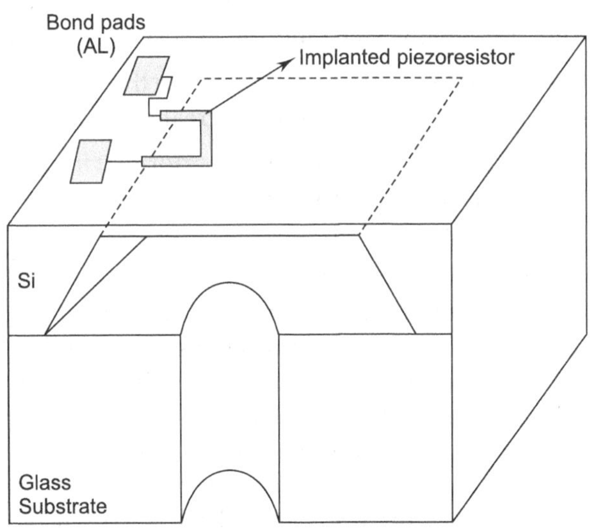

Fig. 7.8 *Simplified drawing of a typical piezoresistive bulk micromachined silicon pressure sensor.*

A membrane is formed using anisotropic wet etching, with piezoresistive elements (only one is shown in Fig. 7.8) to detect the membrane deflection. To maximize their resistance change, the piezoresistive elements, or piezoresistors are positioned at the locations of highest stress. We know that, the locations of highest stress were shown to be along the edges of the membrane at the center of each side. By locating the piezoresistors at these points, the resistance change is maximized, leading to optimal sensitivity of the sensor.

7.5.1 The Sensor Fabrication Process

The sensor fabrication process will define boron piezoresistors implanted into an n-type silicon $\langle 100 \rangle$ substrate, and the membrane will be formed using a timed etch technique. The process involves five photolithography steps, plus dielectric etching and metallization steps. Fig. 7.9 outlines the fabrication process.

Step 2. The first lithography step, defines the regions where the boron implantation in step 3 will done.

Step 3. Boron$^+$ ion implantation. The boron doped piezoresistive elements form the sensing elements for the pressure sensor. The lithography alignment must be done precisely relative to the crystal orientation of the wafer such that the piezoresistors are along the $\langle 110 \rangle$ directions.

Step 4. After the ion implantation and resist removal in Step 3, the implant is activated and 5000 Å of silicon nitride is deposited in Step 4. Si_3N_4 is used instead of SiO_2 because the etch rate of nitride in KOH at 60°C is nearly zero compared with the SiO_2 rate. If SiO_2 is used, a film of thickness of more than 2 μm would be needed to withstand etching of 400 μm of Si.

Step 5. In preparation for making the metal contacts to the piezoresistors, resist openings are created in the Step 5. Lithography process.

Step 6. This step involves etching the exposed Si_3N_4, followed by resist removal.

Step 7. It is the sputter deposition of Al over the entire wafer surface, contacting the Si piezoresistors only in the contact areas created in Step 6.

Step 8. The third lithography process is performed in step 8, where the Al is patterned and removed in the unwanted areas. Processing now moves to the backside of the wafer for membrane formation.

Step 9. It is the fourth lithography step, where the Si_3N_4 on the wafer backside is removed in the appropriate area. The mask alignment for this process is somewhat different than standard lithographic processing. The etch for membrane formation starts from the backside of the wafer, so the opening in the silicon nitride must be oriented properly relative to the piezoresistors already defined on the front side of the wafer. If not, the piezoresistors will not be positioned at the locations of maximum stress, and may not even be on the membrane. This alignment is called a backside alignment step, since the pattern of the backside of the wafer must be aligned with the pattern on the front side (top of the wafer).

Step 10. At this step the membrane is formed using a KOH anisotropic etch. Since the etching process will be stopped at a time determined by the etch rate, the wafer thickness, the desired membrane thickness, the etch rate and wafer thickness must be known precisely. The final step is removal of the Si_3N_4 from the backside of the wafer.

Step 11. Silicon Dioxide Etching :

Step 12. Formation of an anodic bonding of glass substrates. At this point the Si processing is finished, but the wafer, containing many sensor elements, is quite fragile due to the thin membranes. A common packaging method for pressure sensor devices is to bond the finished Si wafer to a glass substrate that contains an array of holes aligned with the membrane. The holes allow the air or other media to reach the membrane. After bonding, a standard dicing saw process can be used to separate the sensor elements for mounting in a commercial package. The pressure input from the package is attached to the hole in glass. Since the pressure medium is conducted to the Si membrane through the hole in the glass. Since the pressure medium is conducted to the Si membrane through the hole in the glass, the bond between the silicon and the glass must be leakproof (hermetic).

This requirement is quite restrictive, considering the lifetime of many pressure sensor applications is measured in years. One particular technique is used for this type of bond, almost to exclusion of all others. Anodic, or electrostatic bonding makes use of the temp-dependent conductivity of certain glasses.

The most commonly used glass corning 7740, contains sodium ions that become mobile at elevated temperature of 400°C. Placing a wafer of 7740 glass in contact with a Si wafer, heating to 400°C, then applying a high dc voltage anywhere from hundreds to a thousand volts as shown in Fig. 7.9, will form a strong band between the Si and glass wafer.

Once the bond is completed, the electric field can be removed and the bonded wafers allowed to cool. If the temperature coefficient of expansion (TCE) of the Si and the glass are significantly different, then a large extrimic stress will develop, bending and perhaps breaking the Si. Glasses such as corning 7740 have been developed specifically to match the TCE of Si, thus avoiding this problem.

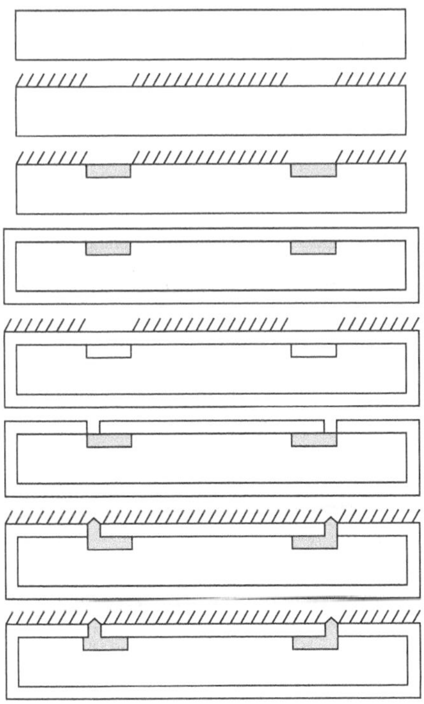

Step 1. $\langle 100 \rangle$ silicon wafer.

Step 2. Mark 1 Lithography process.

Step 3. B^+ ion implantation.

Step 4. Anneal and oxidation.

Step 5. Mark 2 lithography.

Step 6. Contact etch.

Step 7. Aluminium deposition.

Step 8. Mark 3 lithography and Al etch.

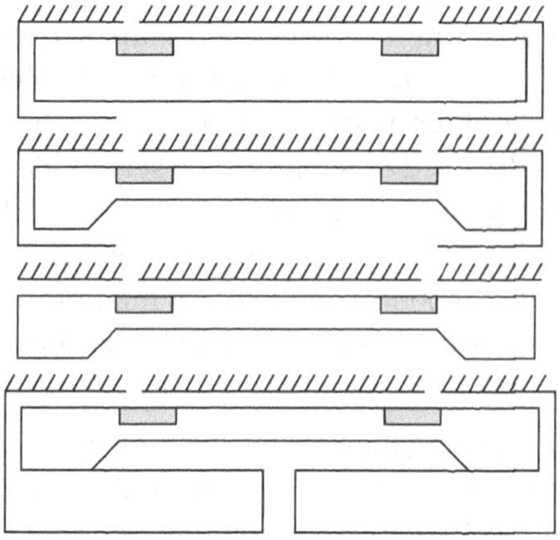

Fig. 7.9 *Bulk micromachining process for a piezoresistive pressure sensor*

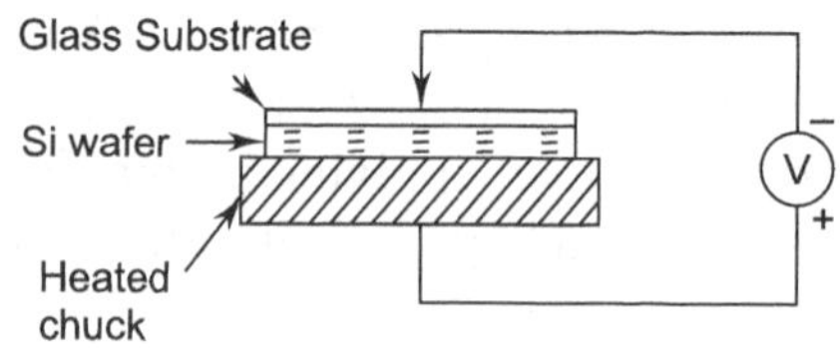

Fig. 7.10 *Typical electrostatic (anodic) bonding setup*

7.5.2 Another Bonding Process

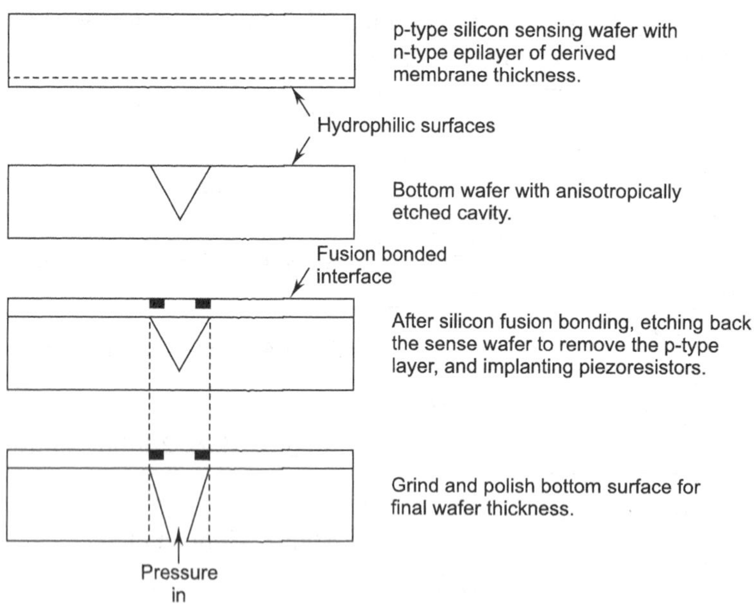

Fig. 7.11 *Fabrication process for Si fusion bonded low pressure sensor*

7.8 SURFACE MICROMACHINING

Surface micromachining is so named because the process takes place on the surface of the wafer, where films used for structural elements are deposited using techniques such as low pressure chemical vapour deposition (LPCVD).

Bulk micromachining is not easily integrated with IC processing. Surface micromachining can more easily be integrated with CMOS processing, allowing signal processing circuitry and MEMS devices to exist on the same chip.

There are two key process steps in surface micromachining : (1) The first is deposition of low stress thin films that can be used for structural elements. (2) The second is the use of a sacrificial layer to allow the structural layer to be detached from the substrate, thus allowing motion of the structural layer. The deposition of controlled stress films is an essential process step for surface micromachining.

Process 1 : Polysilicon deposited by LPCVD is by far the most common MEMS structural material. Standard LPCVD silicon nitride is deposited at temperatures over 800°C using dichlorosilane (DeS) and NH_3 in a flow ratio of approximately 1.5 (Des : NH_4). This process yields a stoichiometric Si_3N_4 film with high tensile stress. By reversing the flow ratio to 5 : 1, a silicon-rich nitride film is deposited and the tensile stress level is considerably lower, less than 50 MPa for the proper deposition conditions. By proper adjustment of the deposition parameters of temperature, pressure, and flow ratio, compressive films are also possible. Post deposition annealing of low-stress nitride films has little effect on the stress. Si_3N_4 has some advantages over polysilicon sensors.

7.7 PRESSURE SENSORS

The unit majority piezoresistive sense elements to detect stress in a thin silicon diaphragm in response to a pressure load.

The basic structure of a piezoresistive pressure sensor consists of four sense elements in a Wheatstone bridge configuration that measure stress within a thin crystalline silicon membrane. The stress is a direct consequence of the membrane deflecting in response to an applied pressure differential across the front and back sides of the sensor. The membrane deflection is typically less than one μm.

The thickness and geometrical dimensions of the membrane affect the sensitivity and, consequently, the pressure range of the sensor.

Devices rated for very low pressures (less than 10 KPa) usually incorporate complex membrane structures, such as central bosses, to concentrate the stresses near the piezoresistive sensors and improve both sensitivity and linearity.

A common design layout on {100} substrates positions the four diffused *p*-type piezoresistors at the points of highest stress, which occur at the center edges of the diaphragm.

Two resistors have their primary axes parallel to the membrane edge, resulting in a decrease in resistance with membrane bending. The other two resistors have their axes perpendicular to the edge, which causes the resistance to increase with the pressure load. Other layouts are also possible including designs to measure shear stress, but the main objective remains to position the resistors in the areas of highest stress concentration in order to maximize the response to applied pressure. It is necessary that the four piezoresistors have identical resistances in the absence of applied pressure. Any mismatch in resistance even one caused by temperature causes an imbalance in the Wheatstone bridge. The resulting output reading is known as zero offset and is undesirable.

Deep diffusions degrade the sensitivity of the piezoresistors by averaging the stress over the depth of the sense element. Shallow diffusions are prone to surface charge effects that can cause long-term drift in the output signal.

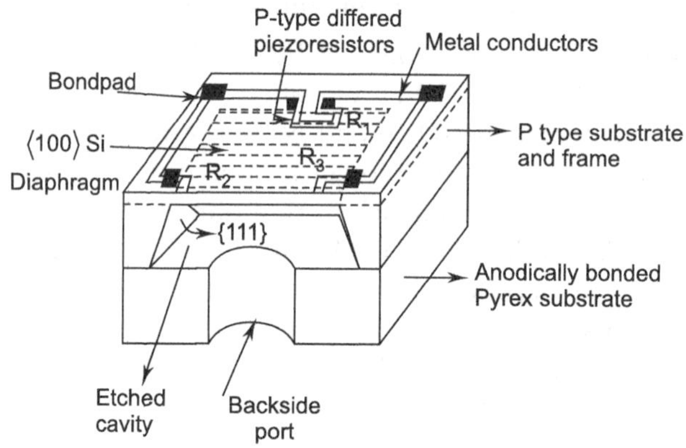

Fig. 7.12 *Pressure sensor with diffused piezoresistive sense element*

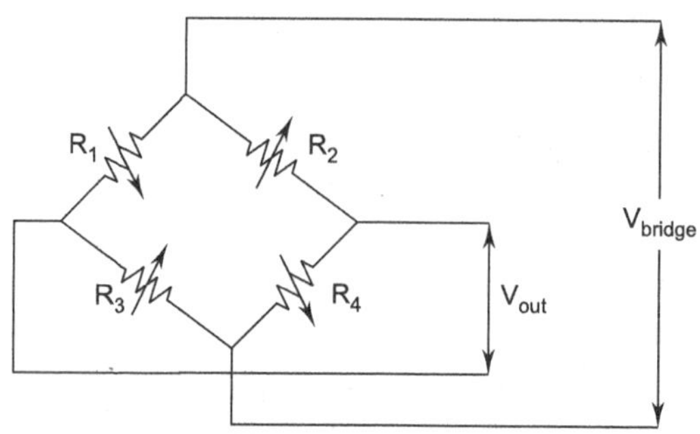

Fig. 7.13 *Four sine elements form a Wheatstone bridge configuration.*

7.7.1 Fabrication Process

Step 1. A *n* type epitaxial layer of Si is grown on a *p*-type ⟨ 100 ⟩ wafer.

Step 2. A thin film preferably stress - free, insulating layer is deposited or grown on the front side of the wafer and a protective Si_3N_4 film is deposited on the back side.

Step 3. The piezoresistive sense elements are formed by locally doping the si *p*-type using the masked implantation of boron, followed by a high temperature diffusion cycle.

Step 4. Etching of the insulator on the front side provides contact openings to the underlying piezoresistors.

Step 5. A metal layer, typically aluminium, is then sputter deposited and patterned in the shape of electrical conductors and bond pads.

Step 6. A square opening is patterned and etched in the Si_3N_4 layer on the back side. Double sided lithography ensures that the backside square is precisely aligned to the sense elements on the front side.

Step 7. At this point, electrical contacts are made to the *p*-type substrate and *n*-type epitaxial layer, and the silicon is electrochemically etched from the backside in a solution of KOH. Naturally, the front side must be protected during the etch.

Step 8. One practical protection method includes coating with wax.

Step 9. The etch stops as soon as the *p*-type Si is completely removed, and the *n*-type layer is exposed. The process forms a membrane with precise thickness defined by the epitaxial layer.

Step 10. Anodic bonding in vaccum of a pyrex glass wafer on the back side produces an absolute pressure sensor that measures the pressure on the front side in reference to the cavity pressure (often vacuum).

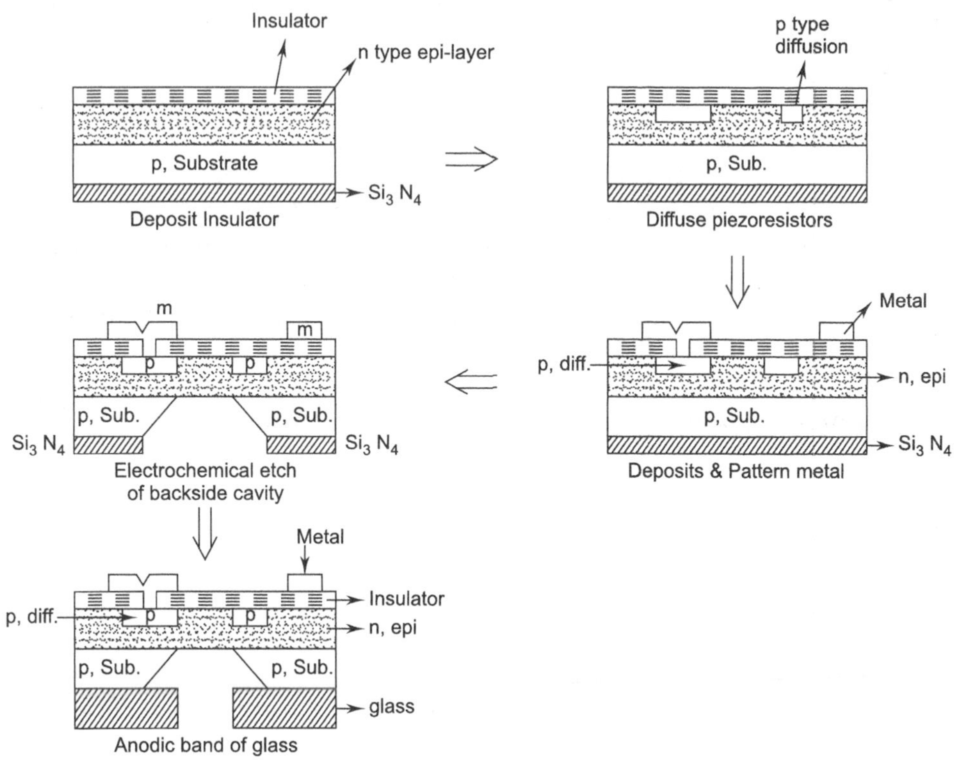

Fig. 7.14 *Fabrication steps for a piezoresistive, gauge or differential bulk micromachined pressure sensor*

A uniform force is applied on the top of the surface of pressure sensor. Higher the thickness, lower the bending. Due to bending, the resistance will change. Initially, we will measure the resistance, when no pressure or force is applied. Now, we again measure the resistance, when pressure is applied. The difference of previous value and present value of resistance is the change in the value of resistance (dR). Bending is significant in the active region.

$$R = \rho \frac{L}{A}.$$

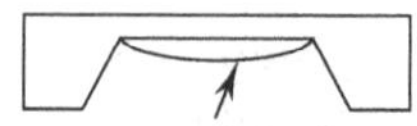

$$\therefore \qquad dR = \frac{d\rho l}{A} + \frac{\rho dl}{A} - \frac{\rho L\, dA}{A^2}$$

$$\text{or,} \qquad \frac{dR}{R} = \frac{d\rho l}{A \times R} + \frac{\rho dl}{A \times R} - \frac{\rho l\, dA}{A^2 \times R}$$

$$\frac{dR}{R} \equiv \frac{d\rho}{A} + \frac{dl}{l} - \frac{dA}{A}.$$

- In case of semiconductor, the change in resistivity that dominates over the change in length and area.
- The amount of fractional range in resistane of semi-conductor, compare to dimension change of metal over application of same pressure.
- The gauge factor of semiconductor is more than metal sensor. Sensitivity depends on gauge factor. So sensitivity of a semiconductor sensor is more than metal sensor.
- In case of metal sensor, the dimension change dominates over semitivity change. The semitivity of *p*-type is more than *n*-type semiconductor.
- The resistivity depends on the band-structure of the semi-conductor.
  The radius of curvature of the ellipse depends on the effective mass. Since effective mass of *p*-type Si is more than effective mass of *n*-type Si.
- When stress is applied, then energy states change, so mobility also change. The conductivity is also changed Diaphragm

7.8 DIFFERENT STEPS IN FABRICATION OF DIAPHRAGM

1. Oxidation
2. Photo-lithography.
3. Selective etching.

$$\frac{\Delta R}{R} = \frac{d\rho}{A} + \frac{\Delta l}{l} - \frac{\Delta D}{D}.$$

$$\frac{\Delta R}{R} = \pi \sigma \text{ where, } \pi = \text{piezo resistive co-efficient.}$$

$$\sigma = \text{stress.}$$

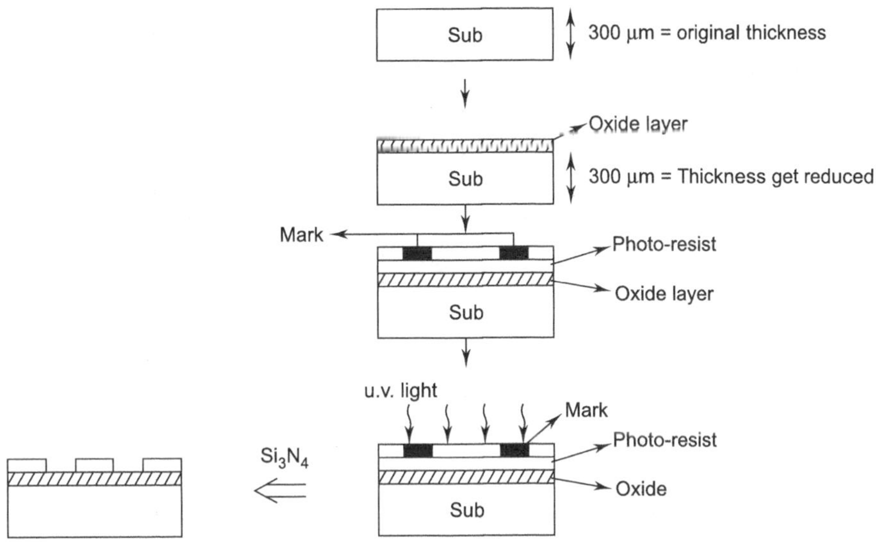

Again, stress $= \sigma = C\epsilon$, C = modulus of elasticity

ϵ = strain

Gauge factor of the si sensor. $\Rightarrow \left(\frac{\Delta R / R}{\epsilon}\right) = \pi C.$

- All the above factors are direction dependent.
- For different direction of the applied force (stress, pressure, etc.), the values of piezo-resistive coefficient (π) is as many as 36, derived from 6×6 matrix. From there, we took only 3 primary values (for si sensor).

During fabrication process, only one value among 3 primary values, is taken. Because the direction of current is fixed, when fabrication is done.

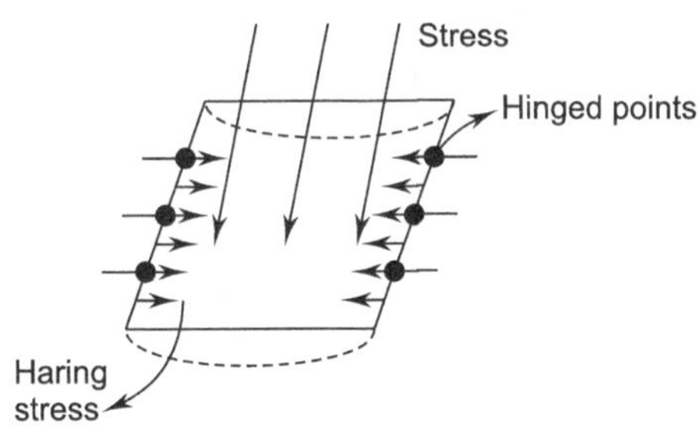

If the sides are hinged, then the sharing forces acts on to play. Sharing stress is perpendicular to flow of current. $\frac{\Delta R}{R}$ is not only contributed by vertical stress, but also by side sharing stress.

Case 1 :

$$\frac{\Delta R}{R} = \pi_l \times P,$$

where π_l = longitudinal piezo-resistive coefficient.

P = applied pressure.

In this case, the direction of the flow of current and pressure applied is same.

This is the Sandwitched configuration.

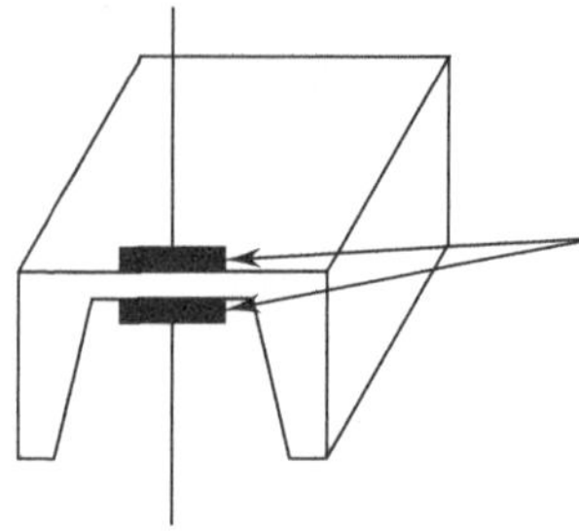

Case 2 :

$$\frac{\Delta R}{R} = \pi_t \times \sigma$$

where $\pi_t \Rightarrow$ transversal piezo-resistive coefficient.

σ = applied stress.

- The sandwitched configuration is rejected as a pressure sensor for low value of π_l and π_t. So, sensitivity is less.
- Range of values of π_l & π_t = 2.7 – 5.8 × 10^{-12} cm^2/dyne.

Factors on which piezo-resistive coefficient depends

(*i*) Depends on the substrate which is used.

(*ii*) Doping concentration of the substrate $\left(\propto \frac{1}{\pi}\right)$.

(*iii*) Temperature.

7.8.1 Lateral Contact Configuration

Stress applied perpendicular to current flow. And sharing stress is formed parallel to direction of current.

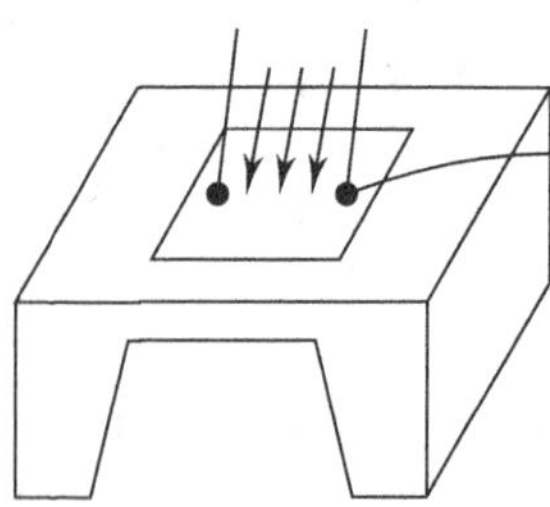

$$\frac{\Delta R}{R} = \pi_t' \, p + \pi_l' \, \sigma$$

values for π_t' and π_l' is roughly, 120×10^{-12} cm^2/dyne.

Fabrication Steps :

1. Measure the dimension of the membrane, which is very much related to stress applied.
2. Start with low doped substrate.

3. Detect, where to place the metal contact.
4. Is selective doping is necessary or not.

Advantage

If we choose proper dimension of the membrane, we can mechanically amplify the sharing stress, 10 times or 20 times or 100 times that of the applied stress.

Let us consider, that pressure is applied on a circular membrane, its circumference is hinged.

$$\sigma_t = \frac{3}{8}\left(\frac{P}{h}\right)^2 \left[a^2(1+\gamma) - r^2(1+3\gamma)\right]$$

$$\sigma_l = \left(\frac{3}{8}\right)\left(\frac{P}{h}\right)^2 \left[a^2(1+\gamma) - r^2(3+\gamma)\right]$$

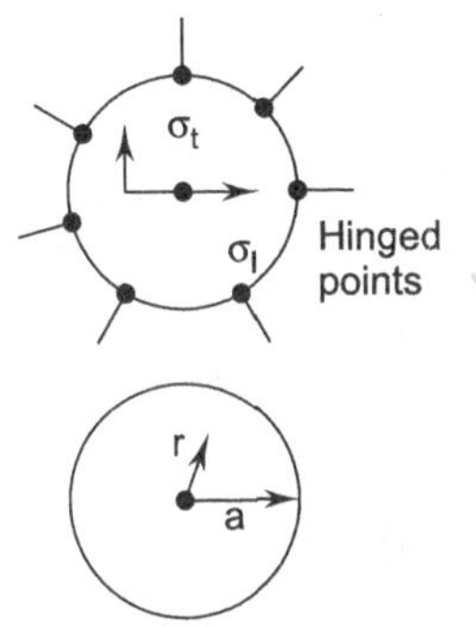

where P is the applied pressure, h is the thickness of the membrane, a = radius, r = radial distance, γ = poison's ratio.

- The above expressions are valid for top surface.
  The resultant stress (σ_r) is minimum near the centre.

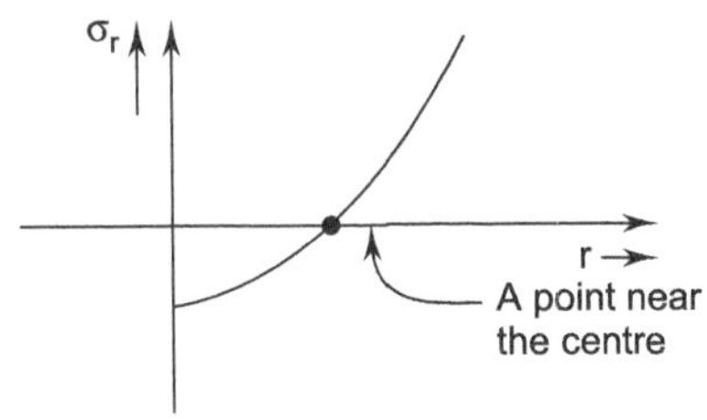

Generated equation for $\sigma_L \Rightarrow$

$$\sigma_L = \frac{3}{4}\left(\frac{P^2 z}{h^3}\right)\left[a^2(1+8) - r^2(3+8)\right]$$

Disadvantage : As, current lines of force will bulge into the membrane.

Nozzles

1. Si_3N_4 deposition on both sides of a (100) wafer.

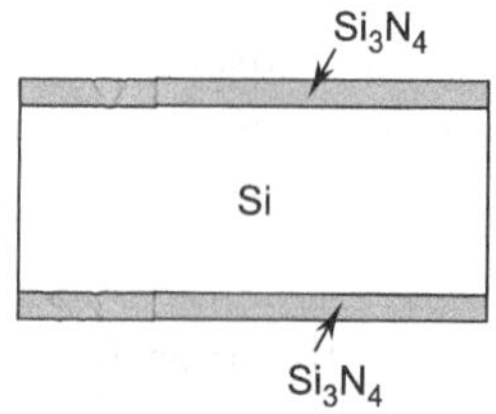

2. Patterning a square in the silicon Nitride layer on the backside.

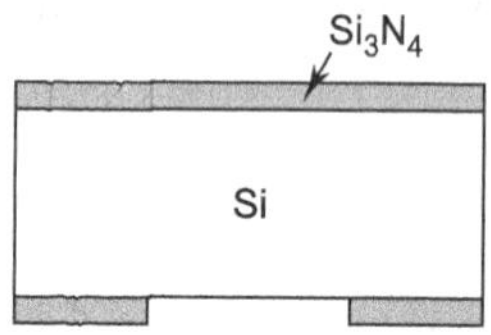

3. Anisotropic Etching in (ROH) or tetramethyle ammonium hydroxide (TMAH) forms a port through the wafer with walls defined by the $\langle 111 \rangle$ planes of silicon.

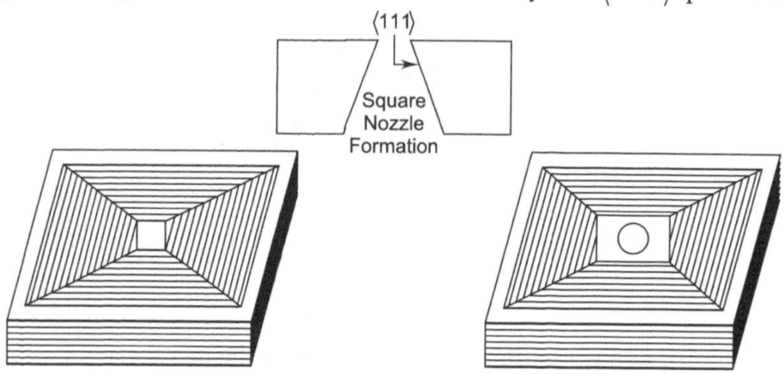

Square Nozzle

Circular Nozzle

1. Growing a *p*-type epitaxial layer of silicon with a high boron concentration ($>1 \times 10^{19}$ cm^{-3}).

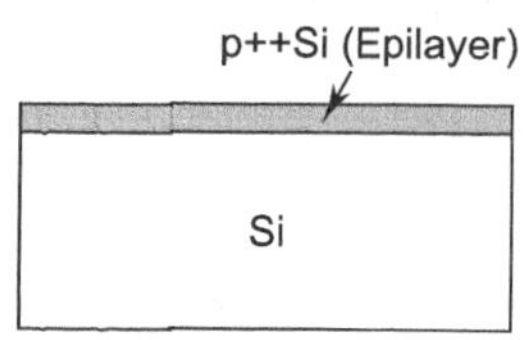

2. The shape of the nozzle is patterned and etched into the *p*-type silicon layer using standard lithography and plasma etching (or RSE).

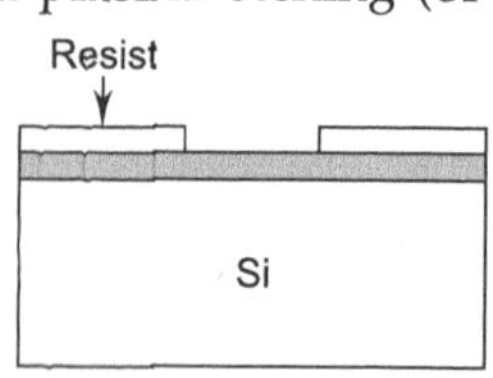

3. A protective layer of Si_3N_4 is deposited on both sides of the wafer and patterned in the shape of a square on the back side.

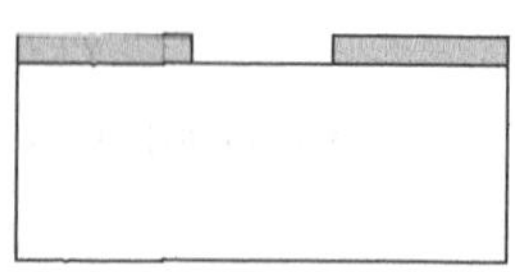

4. Double-sided lithography provides accurate alignment between the nozzle opening and the square on the back side.

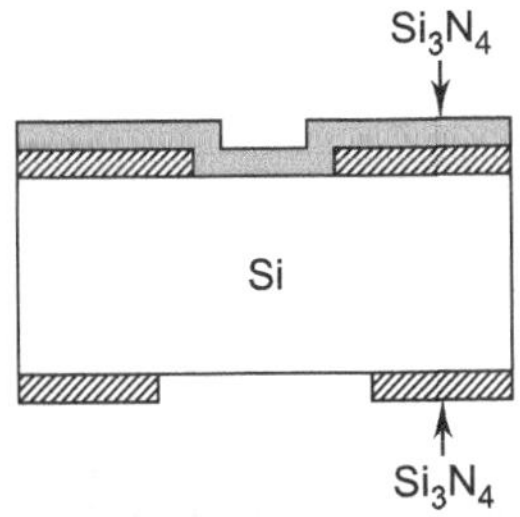

5. The fabrication is complete with the anisotropic etching of the silicon from the back side using KOH or TMAH.

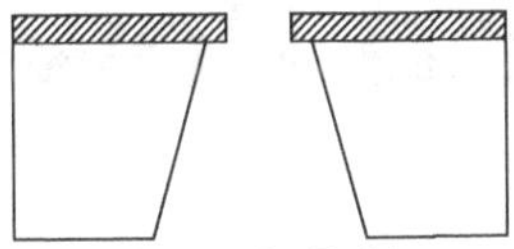

Anisotropic Etching.

The *p*-type layer acts as an etch stop, thus preserving the shape of the nozzle.

These nozzles are oriented perpendicular to the surface of the wafer and are referred to as top shooters or roof shooters in the inkjet field. Nozzles oriented parallel to the wafer surface are termed side shooters.

7.8.2 Moss Flow Meter

If a liquid flow on the hot resistor, then heat loss is occurred in resistor. This gives measurement of temperature of temp.

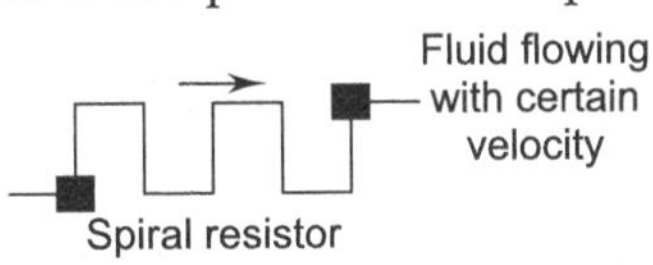

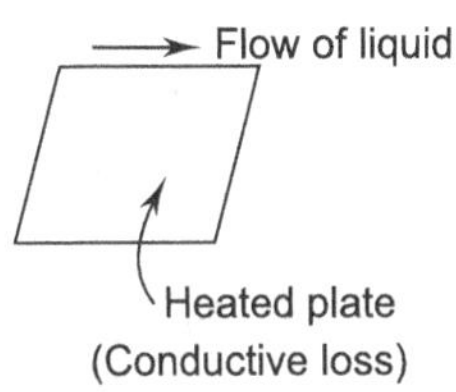

$$V = IR$$
$$\text{Heat} = I^2Rt.$$

Temperature significantly changes due to heat loss, so resistance of the Si increases.

$$\therefore \qquad \Delta T_T = \Delta T\big|_{\text{fluid flow}} + \Delta T\big|_{\text{self heating}}.$$

Growing an insulator & MEMS structure is done to maintain uniform temperature, throughout the heated plate.

QUESTIONS

1. Describe the salient features of a micro-electromechanical system.
2. What is the role of anisotropic etching in bulk micromachining ?
3. Name some materials that can be used as sacrificial layer in surface micromachining.
4. Describe the effect of dopant in silicon bulk micromachining.
5. What is meant by stiction in surface micromachining ?

8

Heterojunction, TED and Power MOSFET

8.1 INTRODUCTION

A heterojunction is formed when two semiconductors with different bandgaps and lattice constants are brought together, usually by epitaxy. Heterojunctions form essential constituents of almost all electronic and optoelectronic devices.

For example, in semiconductor lasers heterojunctions provide both optical and carrier confinement.

In a bipolar junction transistor, the incorporation of a heterojunction drastically improves the emitter junction efficiency and the current gain of the device.

A heterojunction is formed by chemical bonding at the interface. In general, the lattice constants of the two semiconductors are different. If the heterojunction is formed by epitaxy, as is usually the case, then there is a misfit bet^{n} the two semi-conductors, A and B.

But in perfect heterostructure, in which the lattice constants of semiconductors A and B are perfectly matched.

The $Al_x\, Ga_{1-x}$ As alloy system, for the entire composition range, $0 \le x \le 1$, is almost perfectly lattice matched to GaAs. It therefore forms a useful and important heterostructure system.

8.2 ENERGY BAND DIAGRAM

In the formation of a heterojunction with a narrow-bandgap material and a wide-bandgap material, the alignment of the bandgap energies is important in determining the characteristics of the junction. Fig. 8.1 shows three possible situations.

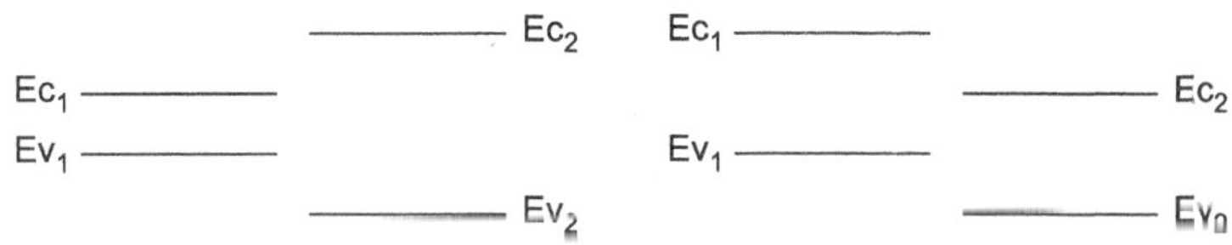

Fig. 8.1

Straddling : The forbidden bandgap of wideband-gap material completely overlaps the band gap of narrow-band gap materials.

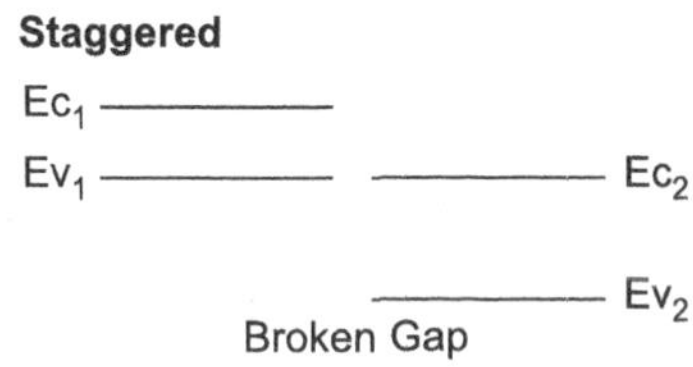

8.3 CLASSIFICATION OF HETEROJUNCTION

There are four basic types of heterojunction :

(*i*) **Anisotype :**
- (*a*) Np
- (*b*) Pn

(*ii*) **Isotype :**
- (*a*) nN
- (*b*) Pp

Capital letter indicates the larger bandgap material.

8.4 TWO-DIMENSIONAL ELECTRON GAS : (ISOTYPE HETEROJUNCTION)

An isotype heterojunction band diagram of an *n*N GaAs - AlGaAs in thermal equilibrium is shown in Fig. 8.2.

χ_1 = electron affinity

ϕ_1 = work function.

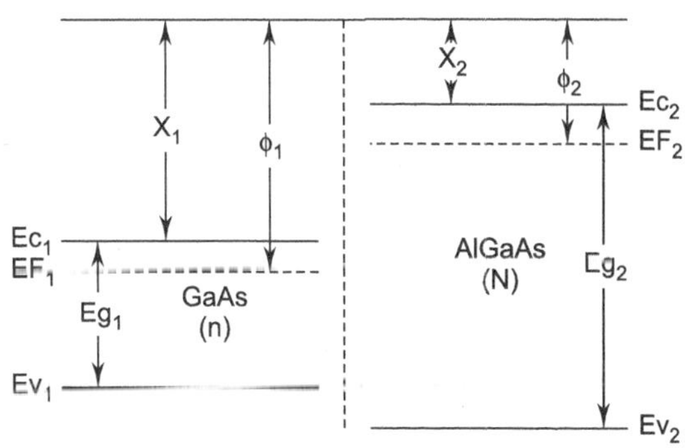

Fig. 8.2 *(a) Two n-type semcionductor with bandgaps Eg_1 & Eg_2 before junction formation*

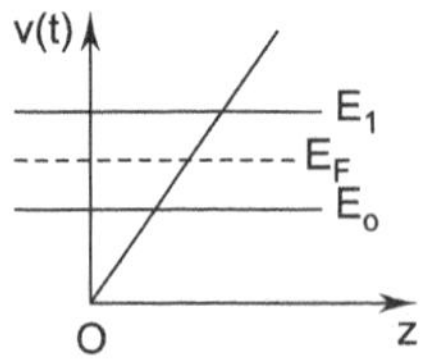

(a) Triangular potential were with desirable electron energies

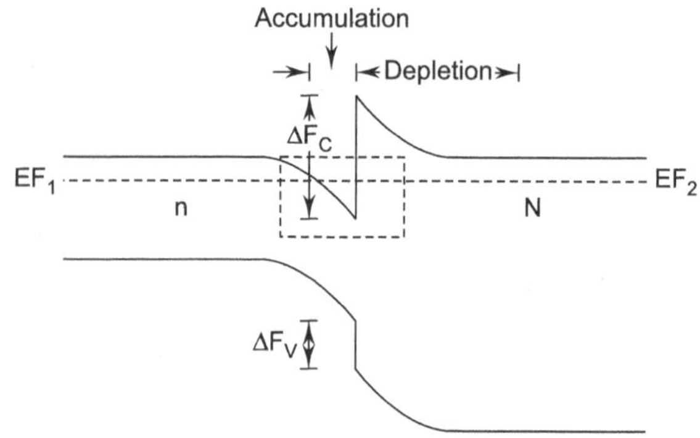

(b) Equilibrium band diagram after junction formation.

Fig. 8.3

The AlGaAs can be moderately to heavily doped *n*-type, while the GaAs can be more lightly doped or even, intrinsic.

To achieve thermal equilibrium, electrons from the wide bandgap AlGaAs flow into the GaAs, forming an acumulation layer of electrons in the potential well adjacent to the interface. This flow of electron occur due to the smaller work function of AlGaAs. The formation of depletion accumulation layer at the interface region is characteristics of heterojunctions.

We also know that, by the basic quantum-mechanical concept, the energy of an electron contained in a potential well is quantized.

The phrase two-dimensional electron gas refers to the condition in which the electrons have quantized energy levels in one spatial direction (perpendicular to the interface), but are free to move in the other two spatial directions.

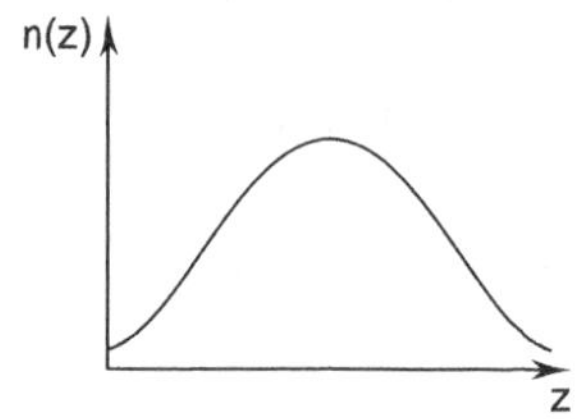

Fig. 8.4 *Electron density in triangular potential well*

For this triangular potential well, we can write,

$$V(x) = e\,E_z\,;\, z > 0$$

$$V(z) = \infty\,;\, z < 0.$$

Remarks :

χ_1 and $\chi_2 \Rightarrow$ electron affinities

ϕ_1 and $\phi_2 \Rightarrow$ work function.

8.5 ANISOTYPE HETEROJUNCTION

A p-N or anisotype heterojunction is shown in Fig. 8.5 given below. Where a small-bandgap p-type semiconductor and a large band-gap N-type semiconductor produce a heterojunction.

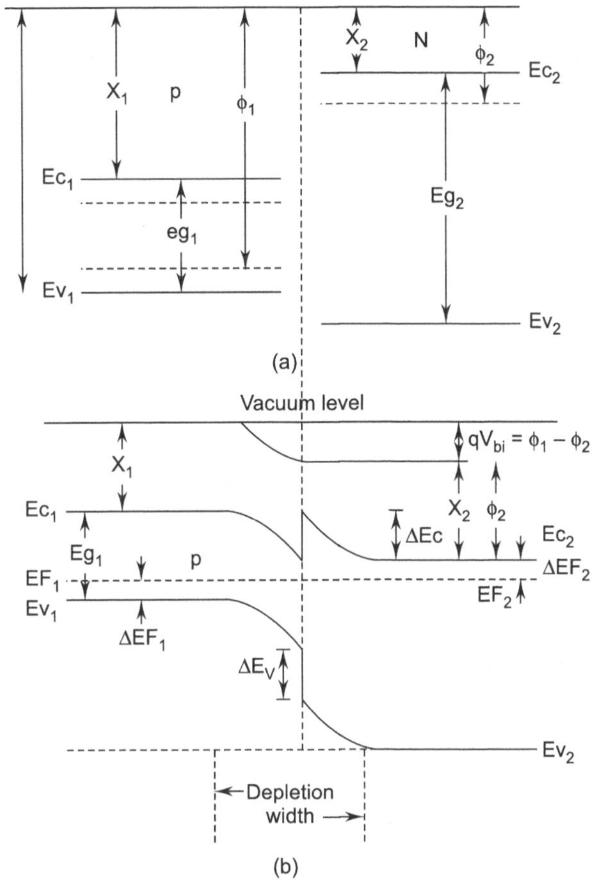

Fig. 8.5 *Band diagrams of p-N anisotype heterojunction (a) before and (b) after contact*

The band bending at the junction is equal to the built-in-potential V_{bi}, which is given by,

$$q V_{bi} = \phi_1 - \phi_2 = (E_{g1} + \chi_1 - \Delta E_{F_1}) - (\chi_2 + \Delta E_{F_2})$$

$$= E_{g1} + \Delta\chi - \Delta E_{F_1} - \Delta E_{F_2} \qquad \text{... (8.1)}$$

According to the Anderson model, the conduction band offset, ΔE_c is given by,

$$\Delta E_c = \chi_1 - \chi_2 = \Delta\chi_1 \qquad \text{... (8.2)}$$

Then the valence band offset is given by,

$$\Delta E_v = (\chi_2 + E_{g2}) - (\chi_1 + E_{g1}) = \Delta E_g - \Delta x \qquad \text{... (8.3)}$$

$$\therefore \quad \Delta E_c + \Delta E_v = \Delta E_g$$

If $\chi_1 = \chi_2$, the offset in the conduction band disappears and $\Delta E_v = \Delta E_g$.

In this case, $q V_{bi} = E_{g1} - \Delta E_{F_1} - \Delta E_{F_2}$

$$= E_{g1} - K_B T \ln\left(\frac{N_c}{n}\right) - K_B T \ln\left(\frac{N_v}{p}\right) \quad \text{... (8.4)}$$

where n & p are the free carrier concentration in the n and p type semiconductors, respectively. Equation (8.4) for the built-in-voltage is identical to that for a homojunction, but there is an important difference, the barrier to hole injection is increased by ($\Delta E_v - \Delta E_g$), compared to a homojunction.

Therefore, the electron injection over the barrier is much more efficient process than hole injection.

∴ The total built-in-voltage, $V_{bi} = V_{bi,\,p} + V_{bi,\,n}$

Where,

$$V_{bi,p} = \frac{t_2 N_2}{\varepsilon_2 N_D + \varepsilon_1 N_A} V_{bi}$$

$$V_{bi,n} = \frac{\varepsilon_1 N_A}{\varepsilon_2 N_D + \varepsilon_1 N_A} V_{bi}$$

The corresponding depletion layer thickness in the p and n regions are,

$$\chi_{dp} = \left[\frac{2\varepsilon_1 \varepsilon_2 \varepsilon_0 N_D}{q N_A (\varepsilon_2 N_D + \varepsilon_1 N_A)} V_{bi}\right]^{1/2}$$

$$\chi_{dn} = \left[\frac{2\varepsilon_1 \varepsilon_2 \varepsilon_0 N_A}{q N_D (\varepsilon_2 N_D + \varepsilon_1 N_A)} V_{bi}\right]^{1/2}$$

The capacitance - voltage relationship is,

$$C(v) = \left[\frac{q \varepsilon_1 \varepsilon_2 \varepsilon_0 N_A N_D}{2(\varepsilon_1 N_A + \varepsilon_2 N_D)(V_{bi} - V)}\right]^{1/2} \quad (\text{Farad. cm}^{-2})$$

8.6 CURRENT-VOLTAGE CHARACTERISTICS

Under applied forward or reverse bias, the change in the band diagram of a heterojunction at the interface region is very similar to that of a homojunction.

Complexities are introduced, however, due to the band offset spikes and the interfacial defects and traps that are present due to lattice mismatch. So, there is some uncertainity concerning the current transport mechanism.

There include thermionic emission over the barrier, interfacial recombination, and tunneling through the spikes and defect states.

Assuming thermionic emission over the heterojunction barrier to be the dominant mechanism, similar to that in a Schottky diode.

The current voltage characteristics can be expressed as,

$$J = \frac{q A T V_{bi}}{K_B} \exp\left[\frac{-q V_{bi}}{K_B T}\right]\left[\exp\left(\frac{q V_a}{K_B T}\right) - 1\right]$$

8.7. TFT : (THIN FILM TRANSISTOR)

The figure of a TFT is given below, where semiconductor (cds) metal and insulator layers are deposited sequentially to form the device.

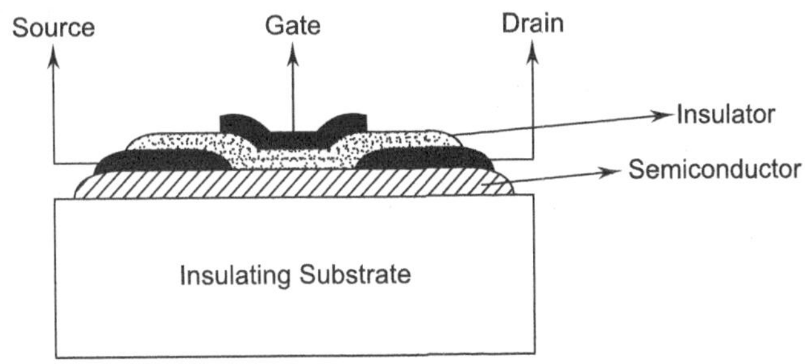

Fig. 8.6

The current voltage characteristics of an enhancement TFT is shown, given below :

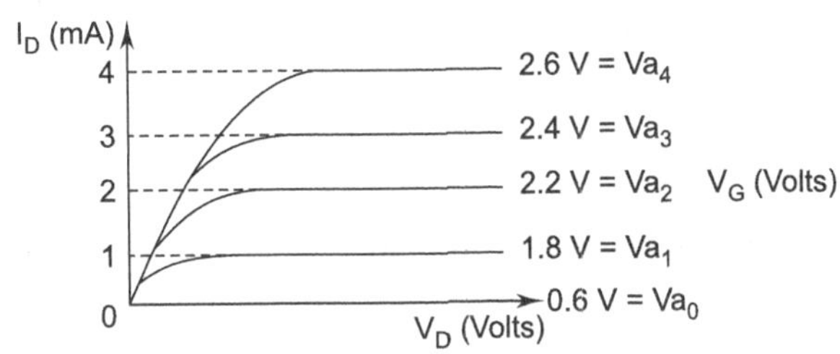

Fig. 8.7

These characteristics are similar to those of a MOSFET. Because the semiconductor layer is formed by deposition, more defects and crystalline imperfections occur in the layer than in the corresponding single-crystal semiconductor, resulting in more complicated transport processes in the TFT.

To improve device performance, reproducibility and reliability, the bulk and interface trap demities must be reduced.

8.8. POWER MOSFET

Power MOSFET has the current handling capability in the ampere range and the drain-to-source blocking voltage may be in the range of 50 to 100 V or even higher.

Advantage

1. The control signal is applied to the gate whose input impedance is extremely large.
2. During the switching between 'ON' and 'OFF' states, the gate current is small, so that relatively large currents can be switched with very small control currents.

Structure

Large currents can be obtained in a MOSFET with a very large channel width. To achieve a large channel width device with good characteristics, power MOSFETS are fabricated with a repetitive pattern of small cells operating in parallel.

To achieve a large blocking voltage, a vertical structure is used.

The vertical channel or VMOS power device is a non-polariar structure that requires a different type of fabrication process.

In this case, a p-base or *p*-'substrate' diffusion is performed over the entire surface followed by the *n*+ source diffusion.

A V-shaped groove is then formed, extending through the n-drift region.

Some anisotropic etchant solution etch the $\langle 111 \rangle$ planes in Si at a much slower rate than the other planes. If (100) oriented silicon is etched through a window at the surface, there chemical etches will create a V-shaped groove. A gate oxide is grown in the V-groove and the metal gate material is deposited. An electron inversion layer is formed in the base or substrate so that current is again essentially a vertrical current between the source and drain. The relatively low doped *n* drift region supports the drain voltage since the depletion region extends mainly into this low-doped region.

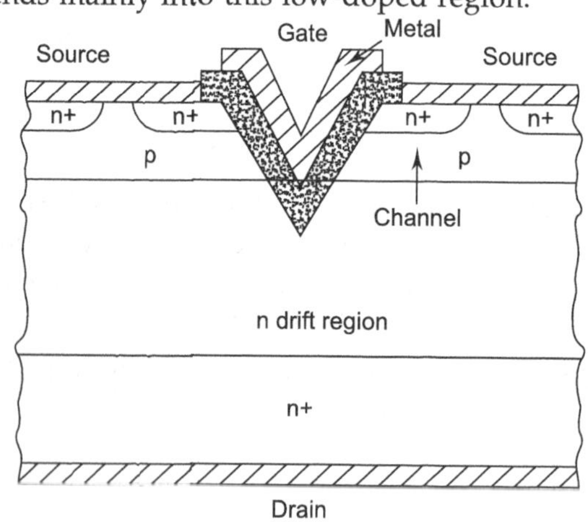

Fig. 8.8

Characteristics

An important parameter of a power MOSFET is the **on resistance** which can be written as,

$$R_{ON} = R_S + R_{CH} + R_D.$$

Where, R_S = Resistance associated with the source contact.

R_{CH} = Channel resistance.

R_D = Resistance associated with the drain contact.

The R_S and R_D resistance values are not necessarily negligible in power MOSFET, since small resistances and high currents can produce considerable power dissipation.

In the linear region of operation, the channel resistance,

$$R_{CH} = \frac{L}{W \mu_n C_{ox} (V_{as} - V_T)}$$

We also know that, the mobility decreases with increasing temperature.

The threshold voltage varies only slightly with temperature so, that, as current in a device increases and produces additional power dissipation, the temperature of the device increases, the carrier mobility decreases and R_{CH} increases, which inherently limits the channel current.

The resistances R_S & R_D are proportional to semiconductor resistivity and so are also inversely proportional to mobility and have the same temperature characteristics as R_{CH}. The typical "ON resistance" characteristics as a function of drain current is shown in given below:

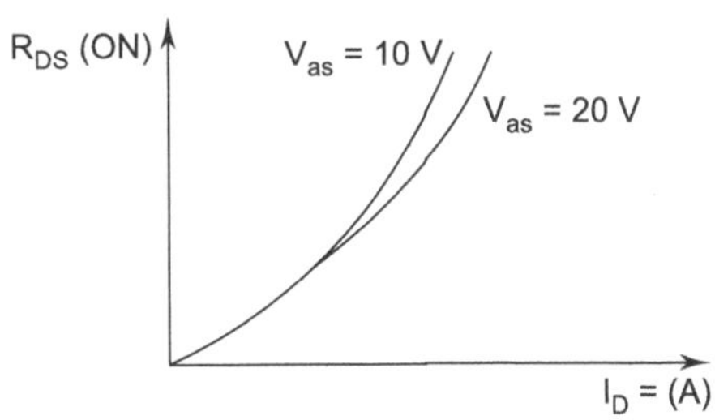

Fig. 8.9 *Typical drain-to-source resistance V_s drain current characteristics of a MOSFET*

The increase in resistance with temperature provides stability for the power MOSFET. If the current in any particular cell begins to increase, the resulting temperature rise will increase the on resistance, thus limiting the current. With this particular characteristics, the total current in a power MOSFET tends to be evenly distributed among the parallel cells, not concentrated, in any single cell, a condition that can cause burn out.

The superior performance characteristics of power MOSFETs are :

(*a*) Faster switching time.

(*b*) No second breakdown.

(*c*) Stable gain.

(*d*) Response time over a wide temperature range.

Safe Operating Area (SOA)

Power MOSFET must operate in a safe operating area (SOA).

SOA is defined by three factors:

(*i*) The maximum drain current, $I_{D_{max}}$

(*ii*) Rated breakdown voltage ($B\ V_{DSS}$)

(*iii*) The maximum power dissipation given by ($P_T = V_{DS}\ I_D$).

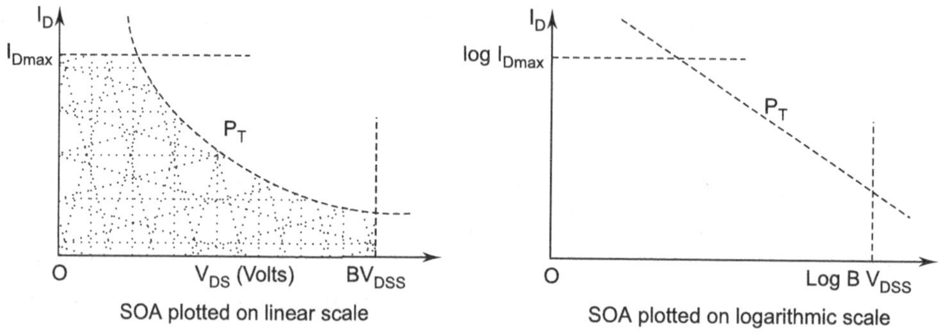

Fig. 8.10

INDEX